MW01453251

Christianity in the 22nd Century

-Jihad, Darwin & Church-ianity-

James Stroud

xulon PRESS

Copyright © 2010 by James Stroud

Christianity in the 22nd Century
-Jihad, Darwin & Church-ianity-
by James Stroud

Printed in the United States of America

ISBN 9781609571047

All rights reserved solely by the author. The author guarantees all contents are original and do not infringe upon the legal rights of any other person or work. No part of this book may be reproduced in any form without the permission of the author. The views expressed in this book are not necessarily those of the publisher.

Unless otherwise indicated, Bible quotations are taken from The New International Version of the Bible. Copyright © 2006 by Zondervan.

Book Edited by Dr. Bonnie C. Harvey

www.xulonpress.com

An actual 10X10 section of the Berlin Wall with Psalm 23 written on it while in Germany.
- May all walls of fascism, hate, and ideology be removed -

I dedicate this book (that began as my Masters thesis) to the persecuted church of Christ which does not usually get any news headlines or recognition, although they are truly advancing the Gospel in ways only God knows; and to the youth of our future – that they may be fully equipped to lead the body of Christ in the future in the name of Christ over doctrine; to my wife Gina – who would support me in any endeavor and is truly a partner; and most importantly, may this simple book bring thanksgiving, praise, and glory to God Almighty in the name of Jesus Christ.

"Despite everything, I believe that people are really good at heart."
Anne Frank

Wisdom (wĭz'dəm)
n.

1) The ability to discern or judge what is true, right, or lasting; insight.
2) Common sense; good judgment.
- The sum of learning through the ages; knowledge.
- Wise teachings of the ancient sages.
3) A wise outlook, plan, or course of action.

"Blessed is the man who finds wisdom, the man who gains understanding, for she is more profitable than silver and yields better returns than gold."
-Proverbs 3:13-14

Introduction

I enjoy movies, so I am going to lead into my introduction by referencing two movies that I feel will open the door to the direction that I am hoping this simple book to take. Most people have seen the 1999 movie The Matrix with Keanu Reeves and Lawrence Fishburne, (if you have not, bear with me as I give you the synopsis). The basic plot of the movie is derived from the philosophical "what if" that most learn about in philosophy 101, (often referred to as the "brain in the vat"). The "what if" is that we are not really here; the entire world we see around us is an illusion, an illusion brought about by all of our experiences, feelings, and every day routines being fed to our minds from a giant computer type system known as the matrix. During the unfolding of the plot, Neo (Keanu Reeves' character) is in a deep search for "truth." During this search for truth, he has a conversation with a person who has a more complete understanding of the "truth" named Morpheus. One of their brief dialogues are as follows:

> **Morpheus:** Let me tell you why you're here. You are here because you know something. What you know you can't explain. You've felt it your entire life. That there is something wrong with the world. You don't know what it is, but it is there like a splinter in your mind. Driving you

mad. It is this feeling that has brought you to me. Do you know what I'm talking about?

Neo: The matrix.

Morpheus: Do you want to know what it is?

Neo: (nods his head yes)

Morpheus: The matrix is everywhere. It is all around us. Even now in this very room. You can see it when you look out your window, or when you turn on your television. You can feel it when you go to work, to church, when you pay your taxes. It is the world that has been pulled over your eyes to blind you from the "truth."

Neo: What "truth?"

Morpheus: That you are a slave Neo. Like everyone else you were born into bondage. Born into a prison that you can not smell or taste or touch; a prison for your mind. Unfortunately no one can be told what the matrix is. You have to see it for yourself.

(Taking out a blue and red pill in each hand) – This is your last chance. After this there is no turning back. If you take the blue pill, the story ends; you wake up in your bed and believe whatever you want to believe. If you take the red pill – you stay in wonderland and I show you how deep the rabbit hole goes. *Remember – all I am offering is the "truth," nothing more.

Jesus of Nazareth, known often as the Christ, repeatedly stated that He was the "truth" itself. Was He? Is He? The story then goes to Morpheus saying that he can not open any doors for Neo; all he can do is show him the door. He then states that all he is offering is a "chance" at the "truth," nothing more. My hope is that we will choose to search and know "the truth" vs. being merely satisfied with the "matrix/status quo" of life, which we all seem to be content living in; but for now, just keep this simple analogy in mind until we conclude this book.

My next movie reference is a less well known movie called "Sliding Doors" with the actress Gwyneth Paltrow. It too has a very interesting plot. Gwyneth's character bumps into a little girl which in turn causes her to miss her train. The movie then diverges into two paths. The first one traces her life by missing the train, while the other one shows her life if she had caught her train. It shows how the smallest of things in our lives can completely and utterly cause it to change course. The series of events that followed her missing the train, showed her life to be beautiful and perfect, while by missing the train, she was thrown into a series of misfortune, suffering, and what would appear to be a doomed existence. Ultimately the movie climaxes with Gwyneth's premature death by making the train before it departed; and in turn, her life which had seemed so destined for failure, began to turn around, and worked out beautifully.

It is now 2009 and I feel forced to contemplate on the realness of this movie's plot in everyone of our lives, and also to that of Christianity. One might say, "yes, for our lives, but not Christianity;" but is it really that unrealistic? Having done my undergraduate work in the field of history, I took a class on Russian history from 1905 to the present, and what I found reaffirmed, was quite alarming. At the beginning of the 20th century, we find Russia as a bastion of the Christian faith. It stood out as one of the promising countries to help in carrying the Christian faith to the rest of the world; an heir to the Byzantine Empire's Orthodox Church, the future was before this colossus of a country. A lesson for the United States and the West to consider is that although the truth of the gospel was widespread in Russia, once it was canonized and ritualized and emptied of the wholeness of its content, it became a hollow shell of pretense and religious hypocrisy, which left the country ripe for revolution. No one could have imagined what means this revolution would use to wage war against its Christian roots.

"The chief nation of the USSR, the Russian Soviet Socialist Republic, was once considered to be among the 'most Christian' nations in the world – a land with a rich, age-old history of churches and monasteries, the wellspring of numerous revered saints and martyrs, iconography and spiritual literature. Yet within less than a year after March 1917, when the last tsar abdicated, a band of militant atheists had seized power; many Russians were looting churches, were mocking religion and religious people unmercifully, and were even murdering priests, monks, and other believers by the thousands. What had happened? How, then, could the Bolsheviks – a small, conspiratorial party determined to smash the Church and root out religion – take over the vast empire in November 1917 and turn it into the world's first atheist state? Even in the midst of revolution, church attendance was high. Today, convinced atheists are still only a fraction of the population. For it was by such techniques of divide and conquer that they were able to subjugate the overwhelmingly Christian Russian Empire. The sad and perplexing story of the Bolsheviks' take-over of a Christian land bears a number of lessons or us today. Churches must be prepared to meet the intellectual challenge of Marxism and other secular ideologies, particularly in the socio-economic sphere (example – Islam); on the one hand, they must not be passive or unaware of political developments; on the other they must not be drawn into facile alliance with latently anti-Christian movements; they must welcome reform from within, but resist manipulation from without; and the list goes on."

-Christian History, Issue 18, 1988, by Andrew Sorokowski, pgs 31 & 34-

What could have Russia been if it had taken a different course a mere 100 years ago? What if Lenin or Stalin would have "not missed their train?" Would the world be forever changed? No one knows for sure. What we do know, is that within the 20th century, Russia destroyed over 90% of their churches, and did everything they could to rip out Christianity by the roots. By the 1950's, over 31 million would lie dead as atheistic communism changed the course of history. Yet I wonder, if just a few pivotal events in history would have been altered, (much like the missing train example), how might things have been? Would Russia have been the beacon of light for the entire world? Would one of those 31 million "numbers" that died, played a huge role in the future? Only God knows, but what we do know, is that a few poor choices within our own personal lives, as well as in our country, church, and Christianity as a whole, can also alter our own course drastically. Like a captain at the helm, one sharp turn of the rudder, can alter the course of the ship drastically. Similarly, the Archbishop of Canterbury and head of the Church of England commented in 2008 that with the rise of Islam in England, he felt that Britain and the church may need to look at adopting parts of Sharia, or Islamic law into the country and possibly the Church. If the principal leader of the Church of England and the head of the Anglican world is willing to make such concessions in favor of fascism in the name of political correctness, compromise and conformity; should we really be surprised that Christianity is in a rapid state of decline in the West, and may face a similar scenario like that of Russia 100 years ago?

A 1940's poster on fighting "Nazism and Fascism;" we likewise will find that we have to stand against the 3 fascist-like ideologies mentioned in this book, for the future of authentic Christianity. (Picture taken by author in Holocaust Memorial Museum)

As I have studied the past through my undergraduate work in history, it reminds me that although we can not change the past, we can learn from it, and use what it teaches us in the present and for the future. Although I am only 31 years of age, it is amazing how the United States has changed in the last 15 years; and how it continues to change. No longer are our sights focused on the betterment of the world; we seem to be worried exclusively about being politically correct and securing our economy so we can continue to base our existence on material possessions, while at the same time, our nation is becoming increasingly hostile to Christ. A very peculiar development as we will discuss. We are seeing the courts filled with lawsuits against a school principal asking for a "moment of silence" being interpreted as forced prayer (after something such as another school shooting), a valedictorian being told that she could not say the word "Christ" in her speech, (and when she did, the school's response of muting her microphone and ending her speech), UCLA informing their graduates that they could not use the word Christ in what they were thankful for, (until lawyers were able to get it overturned), crosses being removed from government cemeteries, ten commandment plaques being forced to be removed from any government buildings, schools continually being limited from following any type of evidence if the conclusions point to some type of creator/intelligent designer; (even though as we'll discuss in Chapter 2, the evidence is becoming more and more convincing for), historians are limited as to where they can take the evidence surrounding Christ, speaking against some of the dangers of radical Islam can result in a hate crime felony, an increased number of bills supporting abortion, and finally an increase in what is now called "specie-ism," which is basically saying that humans are no more special than a rat or any other animal.

And the supposed "Christians" in the West, (which are "supposed" to number 75% of the United States), have for the most part, done nothing... The "writing is on the wall," so to speak, Christianity will be only a shell within the United States' future by the 22nd Century if we continue to do what we have been doing. Before you throw this book down in disgust, I simply want to ask you to research for yourself the topics found in this very basic book, and let me know if anything in here, (from science, history, theology, philosophy, and society), is false. If it is, please go to one of my websites or amazon.com and post the errors and I will edit any future copies. The point of this simple book is to touch on the basics of the past that we could be doomed to repeat - where the world stands presently as of 2009, and where we are bound to be as we enter the 22nd Century if we do not do something now; and to touch on what we can do at present to "not miss the train."

This simple book is written in a manner that should be able to intrigue the specialist, but also be applicable to the layman alike. A full treatment of Christian persecution/history, apologetics/theology, and our reaction to them all, is far beyond the scope of this book, nor is it my purpose. The purpose of this book is to provide a solid backing of these points, and provide a concise source that can quickly be referenced in helping you better understand the Christian faith, and expressing the three points I highlight to others. It is my sincere hope to give a faithful but stern testimony to believer, unbeliever, and skeptic alike, and that after reading this small book, each reader should be able to:

1) Follow the forgotten historic path of Christianity that did not end in success and material advantage, but in death and destruction; the discovery should in turn instill a sense of integrity and honor in both believer and skeptic. (Hebrew 13:3 – Remember and support our persecuted Christian brethren)

2) See why the Christian faith is viable; and that the believer must "always be prepared to give an account for their faith to anyone that asks them." 1 Peter 3:15 (Moreover, work to insure that our schools and government are truly following the "evidence" and not their own secular agenda, when it comes to our educational systems.)

3) To show briefly where the evidence points in biology, astronomy, philosophy, and history; and once again show why theism, and moreover "real" Christianity, (in the words of William Wilberforce), is a viable option, that anyone who measures all the evidence will have to admit does make irrefutable sense.

4) To show the Christian that "Church-iantiy" is different than "Christ-ianity," and that the majority of us who call each other Christians, are judgmental,

hateful, and not following in the real footsteps of Christ; hopefully to reflect on who we really see in the mirror.
5) Finally to equip and motivate you with the basics to do something in your world, not for profit or gain, but in discipleship to Jesus of Nazareth.

I hope these three simple chapters will at least instill in you, a spark of what Christianity is, and what it can be in the 22nd century, what its biggest obstacles are, and how we can overcome them.

2009

As I am presently writing this section, it is November, 2009. It is quite ironic that this year marks the 200th anniversary of Charles Darwin, and this month the 150th anniversary of his book: Origin of Species. Much less remembered, though, is that this year also marks the 100th anniversary of Richard Wurmbrand. A Romanian Jewish Christian pastor, who spent 14 years in a communist prison for sharing his faith to an atheist government; how ironic that both Darwin's work on evolution and Pastor Wurmbrand's message on persevering through persecution are both in many way completely related. In 2005 the United States government ruled that Intelligent Design, (that evidence is pointing more and more to the fact that there must be an intelligent designer), can not be taught in the same light as the religion of Darwinism in public schools, (see Dover vs. Kitzmiller). It is ironic, that Darwinism + Marxism were two of the combinations that led to the Communist state of Russia in the first place; and now we celebrate the 200th birthday of Darwin and the 100th of Wurmbrand. It is ironic too, that while Darwin started life as a believer, and ended as by most accounts, an agnostic; Wurmbrand, in his own words, began his journey as an atheist, until he made a simple prayer for truth:

> "One day, being a very convinced atheist, I prayed to God. My prayer was something like this: 'God, I know surely that You do not exist. But if perchance You exist, which I contest, it is not my duty to believe in You; it is Your duty to reveal Yourself to me.' I was an atheist, but atheism did not give peace to my heart." [145]

Charles Darwin went to a Christian college, but ended up taking the controversial theory of evolution to a larger naturalistic level than even he could have ever dreamed. The theory of evolution would play a huge part in the future life of Richard Wurmbrand as well, due to the fact that Lenin, Stalin, and Hitler (just to name three), would be largely influenced by this naturalistic explanation that allowed them to push God out of the picture. As ironic as it was, we "almost" have a case of preacher turned naturalist, and an atheist turned preacher; it seems that once again Christians will face an ever growing tide of persecution for their faith, as an ideology known as atheism/Darwinism/naturalism continues to grow and attack Christianity once again.

> "Wherever people know how to write, they have a holy book. Atheists, too, have one – it is called *The Atheist's Handbook*. It was first published in 1961 by Moscow's Academy of Science, (the State publishing house for political science). This book, which has been reprinted many times, has been translated into many languages and widely distributed in other socialist countries. From the primary grades through college, on radio and television, in films and at atheistic rallies, the ideas contained in this book are propagated. The primary purpose of the book is to show that there is no God."
> (*The Answer to the Atheist's Handbook*, Richard Wurmbrand pg 5)

The Atheist's Handbook was drilled into children at school and adults at home. Ironic that Wurmbrand would suffer for 14 years in these prisons, and find his faith increased; he then wrote the book quoted above (The Answer to the Atheist's Handbook), where he showed how weak their arguments were. Another bit of irony is that once again, we are in a generation where militant atheists such as Richard Dawkins, Sam Harris, and Christopher Hitchens, etc., are each writing their own "Atheist Handbook," yet again. Again, we are seeing our schools enforce the teaching of naturalism, (which excludes any mention of the supernatural); is this teaching really that different than what Richard Wurmbrand experienced in the 1940s-60s?

> *"In all the Christian countries of the West, atheism has full liberty for its propaganda. Christianity has not the slightest reason to fear it. In free debate, only Christianity can win. Imagine two rooms separated from each other by a thick curtain. In the one darkness reigns, the other is lighted by a candle. If the curtain is withdrawn, it is not darkness that prevails. Darkness cannot overcome the light from the candle, because it is not energy. It is the absence of light. Only light, being energy, can prevail. Thus, the room that was in darkness becomes visible, transformed by the burning candle. Christians have not feared prisons nor the implements of torture. Neither do we fear atheist books. In the struggle of ideas, the final victory can only be ours."*
> (The Answer to the Atheist's Handbook, Richard Wurmbrand pg 8)

Christ promised that if we follow Him, we would be hated. So we should not be surprised when we see throughout

Christianity's history, governments, atheists, Islam, etc., have sought to destroy it, but all have failed, so take courage; but I do sincerely hope and pray you do not remain on the sidelines. It would have been too easy for Richard Wurmbrand to have denied Christ and be freed from prison; for an average of 150,000 Christian Martyrs per year in Islamic countries to have renounced their faith in Christ; for the William Wilberforce's, or Martin Luther King Juniors of the world to have done nothing; but they all truly lived with Christ inside of them, and we can too.

We seem to live at an odd point in history; a point that is almost retractable back to a possible feeling of what early Christians may have felt like – being looked down upon for being Christian throughout the world, except for the Americas, which along with Europe, seem to be slowly but surely turning down the road to a point where the world is once again hostile to the Christian faith.

In his crucial, (yet unfinished), book ETHICS, Dietrich Bonhoeffer seems to have been sadly accurate in his 1930's projection of the direction he was afraid the west might be turning concerning Christianity:

> *"The west is becoming hostile towards Christ. This is the peculiar situation of our time, and it is genuine decay. The task of the church is without parallel. The corpus christianum is broken asunder. The courpus Christi confronts a hostile world. The world has known Christ and has turned its back on Him, and it is to this world that the Church must now prove that Christ is the living Lord. The more central the message of the church, the greater now will be its effectiveness."*
> *(ETHICS by Dietrich Bonhoeffer, pg 109)*

Why this growing hatred of Christ from the world? Should we worry? What should we do? Once again – we shouldn't be surprised. Christ was very clear when he repeatedly said, "If they persecute Me, they will persecute you." Eleven of the twelve Apostles were martyred after all. Early Christians such as Justin Martyr said in his apology, that as Christians, we know we are to suffer and die, like it was just a given fact of life. Many Christians throughout the Middle East, Asia, and Africa are still persecuted and killed for following Christ; and yet I wonder if we in the United States could even sustain minor persecution such as losing a car or our jobs for our faith? Why in the Christian United States is it becoming almost a bad word to mention the name of Christ? Some will obviously say: "Who cares?" Well, let's first look at why you should care before anything else in your life.

People who shrug their shoulders and say, "What difference does it make whether God exists or not?" simply reveal that they have not fully grasped the impact of this question, or its ramifications. As philosopher William Lane Craig puts it: "If God does not exist, then life is ultimately meaningless. If your life is doomed to end in death, then ultimately it does not matter how you live. In the end it makes no ultimate difference whether you existed or not. Sure, your life might have a relative significance in that you influenced others or affected the course of history. But ultimately mankind is doomed to perish in the heat death of the universe. Ultimately it makes no difference who you are or what you do. Your life is inconsequential."[1]

Craig is accurate in his conclusions. Think about it. If you're a billionaire, own a country, find the cure to cancer, and solve global warning – what difference does any of it make outside of God? In 100 years who will even know your name, and even if they do, the sun will eventually expire/ super nova thus ultimately destroying our planet, (if we don't do it ourselves); so with that, the human race will no longer even exist, so outside of God, there is not ultimate meaning or purpose to life. That is why one should not simply shrug their shoulders and say "what difference does it make?" It makes all the difference, and in all honesty, the only true difference.

With that morbid picture out of the way – let's discuss the wonderful news. God most definitely does exist and one does not have to be a theologian to discover this. So we are going to briefly review why we should be reviewing someone like Richard Wurmbrand's life more thoroughly than Charles Darwin's, as we celebrate both birthdates this year; we will also see why there is so much hope in light of the darkness surrounding Christianity in the guise of militant atheism, fascist Islam, and internal strife and compromise within the ranks of our own churches as we slowly approach the not so distant 22nd century.

How ironic is it that one person in all of human history sticks out in every continent? One person recognized as the Essence of God from Mt. Sinai, who then dwelled among the people in the Jewish Tabernacle, then in King Solomon's Temple, and finally in the virgin Mary as the prophets foretold; He was born as the sinless Savior of the world in the temple of flesh and blood – Jesus of Nazareth; the promised Messiah of the Old and New Testaments; the same person told about in the Muslim's Holy Koran, who will come at the end of the age; the person who is considered by most Buddhists to have reached their ultimate stage of enlightenment known as Nirvana; the same person considered by many Hindu to be a God; the same person who has divided human history, and the only religion that says you do not have to earn anything, but simply accept the free gift of his sacrifice for the payment of sins – the person known collectively as Jesus Christ.

> *"The near-universal appeal and attraction which Jesus Christ has evoked even outside Christianity confirms the claim that only he can truly unite people of all cultures. It is notable that the three largest non-Christian religions all have sought to come to terms with Jesus in some way. Islam, which numbers roughly a billion people, views Jesus as a great prophet and miracle-worker. Hinduism, numbering roughly 750 million, often views Jesus as an avatar of Vishnu — one of many incarnations of one of the many Hindu forms of God. Buddhism, which accounts for about 300 million people of the world, typically regards Jesus as an enlightened one for the West. What these religions unwittingly attest by extending such honors to Jesus is that he is the one religious figure in history that simply cannot be ignored."* [5]

I pray that as 2009 also marks the 20th anniversary since the Berlin Wall of atheistic Russian communism was torn down, that we may also, (while keeping Christ in the center), work to build bridges with all persons, to tear down the walls of extreme secularism that lead to an abandonment of God, and thus lead to the suppression in academia of the truths that point to a creator, the Islamic wall of Fascism that suppresses its own members, blocks internet usages, kills moderate Muslims (or anyone else), that challenge what is wrong or not logical, and finally to tear down the wall of Christian hypocrisy; to focus on Christ and not our own selfishness. (See appendixes)

May this book help to shed light on some of these neglected subjects, while we still have an adequate amount of time to affect the future as we approach the 22nd century;

and moreover, may this simple book bring glory and focus to Christ Jesus. – Amen

```
        Authentic Faith
              /\
             /  \
            / CHRIST \
           /_____\
      Persecution   Apologetics
```

This triangle represents the direction and flow this simple book hopes to take, with each point on the triangle representing a chapter in the book.

Matthew 25:14-29 (New International Version)

The Parable of the Talents

[14] "Again, it will be like a man going on a journey, who called his servants and entrusted his property to them. [15] To one he gave five talents of money, to another two talents, and to another one talent, each according to his ability. Then he went on his journey. [16] The man who had received the five talents went at once and put his money to work and gained five more. [17] So also, the one with the two talents gained two more. [18] But the man who had received the one talent went off, dug a hole in the ground and hid his master's money.

[19] "After a long time the master of those servants returned and settled accounts with them. [20] The man who had received the five talents brought the other five. 'Master,' he said, 'you entrusted me with five talents. See, I have gained five more.'

²¹"His master replied, 'Well done, good and faithful servant! You have been faithful with a few things; I will put you in charge of many things. Come and share your master's happiness!'
²²"The man with the two talents also came. 'Master,' he said, 'you entrusted me with two talents; see, I have gained two more.'

²³"His master replied, 'Well done, good and faithful servant! You have been faithful with a few things; I will put you in charge of many things. Come and share your master's happiness!'

²⁴"Then the man who had received the one talent came. 'Master,' he said, 'I knew that you are a hard man, harvesting where you have not sown and gathering where you have not scattered seed. ²⁵So I was afraid and went out and hid your talent in the ground. See, here is what belongs to you.'

²⁶"His master replied, 'You wicked, lazy servant! So you knew that I harvest where I have not sown and gather where I have not scattered seed? ²⁷Well then, you should have put my money on deposit with the bankers, so that when I returned I would have received it back with interest.

²⁸" 'Take the talent from him and give it to the one who has the ten talents. ²⁹For everyone who has will be given more, and he will have an abundance. Whoever does not have, even what he has will be taken from him.

What will you do with the coin you have been given in life?

CHAPTER I

Did Christian persecution end in Rome?

> *"Islam is willing to destroy for the sake of its ideology. I want to suggest that the choice we face is really not between religion and secular atheism, as Sam Harris, Richard Dawkins, Christopher Hitchens, and others have positioned it. Secularism simply does not have the sustaining or moral power to stop Islam. Even now, Europe is demonstrating that its secular worldview cannot stand against the onslaught of Islam and is already in demise. In the end, America's choice will be between Islam and Jesus Christ. History will prove before long the truth of this contention..."* [132]
> - Ravi Zacharias, The End of Reason, pgs 126-27

Wow... This quote by a well educated/cultured Ravi Zacharias seems quite alarming, blunt, not politically correct, and straight to the point. Is he simply stating an opinion, or a statement based on facts? It is true that as of 2009, approximately 90% of all violence in the world has fascist Islam on at least one side of the conflict; it is also true that fascist Islamic countries do not allow freedom of religion or equal rights for women; it is also true that fascist Islam has accounted for almost 10,000,000 Christian martyrs in its history (on the conservative side), as well as over 200,000,000 million deaths by some estimates (some estimate more, some less), of Christian and non-Christian alike in its history; so there may be some truth to Ravi's statement. I do have a wonderful Muslim friend whom I will talk more about in the third chapter of this book, but while she is a devout Muslim living in the United States, she readily admits that a true reading of the Koran (Islam's holy book), she would be killed many times over. She drives a car, has a job, doesn't wear a head covering, she talks to me (a male that is not related to her), and she has found joy in the simplicity of Jesus of Nazareth (even though she is still Muslim as we'll discuss more in chapter 3). All of these are worthy of death in the vast majority of Islamic countries and in a true interpretation of Islam.

> *"Fight those who believe not in Allah nor the Last Day, nor hold that forbidden which hath been forbidden by Allah and His Messenger, nor acknowledge the Religion of Truth, from among the People of the Book (Christians), until they pay the Jizya (non-Muslim tax) with willing submission, and feel themselves subdued." (Koran – Surah 9:29)*

Ayann Hirsi Ali describes her ordeal as a woman in a truly Islamic world (that my friend Nouran has thankfully been exempt from), in her books The Caged Virgin and Infidel. Ayann burst into the news with the brutal murder of Theo Van Gogh by an Islamist who said that she would be next. Although she had escaped the clutches of Islam and was living in the Netherlands, the Dutch parliament in an act of political correctness to appease Islam, revoked her citizenship. She survived beatings, female mutilation, constant battles and civil wars in the name of Islam, to tell her story. (www.AyannHirsiAli.org)

In the February 2010 copy of International Christian Concern newsletter, Ayaan has commented that "the more Islam you practice, the more inhuman your practices will be. So, people we call radical Muslims – that is a euphemism for practicing Islam pure and simple. ...For example Prophet Muhammad married a nine-year-old girl, and all Muslim men who want to marry an underage girl will invoke Muhammad and ignore the constitution of the country they live." She goes on to express her concerns that even though she now lives in the United States, she still has to have protection for her life, because she left the ideology of Islam.[156]

While I am in no way "bashing Muslims," many of whom are my friends and we have discussed this subject in detail, we must recognize the facts as Joel Richardson lays them out:

> *"Islam is the fastest growing religion in the United States, Canada, and Europe. The annual growth rate of Islam in the U.S. is approximately 4 percent, but there are strong reasons to believe that it may have risen to as high as 8 percent over the past few years. Every year, tens of thousands of Americans convert to Islam. Prior to 2001 most reports have the number at roughly twenty-five thousand American converts to Islam per year. This may not sound like much, but this yearly figure, according to some Muslim American clerics, have quadrupled since 9/11. That's right: since 9/11 the number of American converts to Islam has skyrocketed. As early as one month after the World Trade Center attacks, the reports began flowing in from mosques all over America."* Joel also ties in the sad reality that: *"The other sad aspect of*

these figures: Over 80% of these American converts to Islam were raised in Christian churches." [157]

Please consider these above statistics as you read the rest of chapter 1. I have a true love for Muslims but not Islam; I hope if anything, this chapter will encourage these same sentiments, and allow us to reflect on how we are showing this love. I am 100% against showing it through political correctness which is illogical and unbiblical, but I am 100% for showing it through genuine relationships and outreach. Paul tells us in Romans 10:14: *"How can they believe in the one of whom they have not heard? And how can they hear without someone preaching to them?" Only 2 percent of the Protestant missionary force is reaching out to the Muslims of the world, who make up practically half of the non-Christian world population.*[159] If you are a Christian, then I would seriously challenge you to reflect on Christ's command to "go and make disciples of all nations, baptizing them in the name of the Father, Son and Holy Spirit." It truly seems that once again the words of Luke 10:3 ring out – "The harvest is plentiful, but the workers are few." There are a number of things you can do that include prayer, support financially organizations that target getting Christian information into Muslim hands (even though it is outlawed by the government in many Islamic countries), and we also must be open to seeing the growing Muslim population around us, and not just ignore them. Through my personal experiences, most Muslims in the United States are just like anyone else, and they are very open to genuine relationship and open dialogue. As we'll discuss briefly in the next chapter, once we reach out with the truths of Christ's atonement and address some of the objections Muslims have, all areas of searching point back to Christ, (this is not just a presupposition – for example the Bible being written over the course of 1,500 years by over 40 authors conveying the same message pointing to Christ with a plethora of witnesses, archaeological evidences, etc., vs. Islam's claim made by one author (Mohammed) who did not know himself if his revelations were accurate at first, and who did not write the Koran, which was written over 600 years after Christ; the open-minded Muslim follower must then begin to reflect on these facts, and the credibility of each). If the Christian will truly engage in loving, open, and prayerfully mediated dialogue with their Muslim Brothers and Sisters, the "truth" speaks for itself. This is why most Islamic countries outlaw any form of evangelism; if they did not have a forced ideology of Islam forced on its people, and they had access to all the facts with freedom of religion; then I estimate that approximately 30-50% of all Muslims would leave Islam within 10 years. (See appendices for more info)

Ravi Zacharias sums this point up well in his book: *"Jesus Among Other Gods – "The teaching of Jesus is clear. No one ought to be compelled to become a Christian. This sets the Christian faith drastically apart from Islam. In no country where the Christian faith is the faith of the majority is it illegal to propagate another faith. There is no country in the world that I know of where the renunciation of one's Christian faith puts one in danger of being hunted down by the powers of state. Yet, there are numerous Islamic countries where it is against the law to publicly proclaim the gospel of Jesus Christ, and where a Muslim who renounces his or her belief in Islam to believe in anything else risks death. Freedom to critique the text of the Koran and the person of Mohammed are prohibited by the laws of blasphemy, and the result is torturous punishment. One must respect the concern of a culture to protect what it deems sacred, but to compel a belief in Jesus Christ is foreign to the gospel, and that is a vital difference. The contrast is all too clear. Jesus' method was to touch the heart of the individual so that he or she responded to Him out of love for Him, rather than from compulsion or control. Contrast this with the practice of Mohammed. However one might wish to interpret* it, the sword and warfare are an intrinsic part of the Islamic faith. Even the best of apologists for Islam acknowledge the use of the sword in Islam but will mitigate it by saying that in each instance it was for defensive purposes. I suggest that the reader read the Koran and the history of Islam for himself to determine whether this was so.

Even the best of Muslim apologists is hard-pressed to navigate around Mohammed's own injunction to kill, illustrated in a verse from the Koran knows as the ayatus-saif, or 'the verse of the sword.'"[160]

> "But when the forbidden months are past, then fight and slay the idolaters wherever ye find them, and take them, and prepare for them each ambush. But if they repent and establish worship and pay the poor due, then leave their way free." Surah 9.5

"The Power of Islam" display from the New York City Museum of Natural History – taken by author's wife.

Gifted speaker, expert on Islam and Sharia Law, Sam Soloman sheds some light on the fascism at the heart of Islam, when he share his own experiences in Ravi Zacharias' book "Beyond Opinion" when he states:

> "Islam cannot be defined as a religion in the Western sense of the word; neither can it be termed as a faith. Muslim scholars state that Islam is an all-encompassing system – a sociopolitical, socio-religious system, as well as socio-economic, socio-educational, legislative, judiciary, and military system governing every aspect of the lives of its adherents, their relationships among themselves, and with those who are non-Muslims. I am sad to say that not only were millions and millions of people killed in those early Islamic conquests, but jihad in real terms has virtually never stopped. Today persecution is rife, and non-Muslims have at best a token freedom for religious survival with severe penalties if they ever attempt to preach overly or attempt to convert Muslims. Because of this climate of fear regarding charges for converting Muslims, when I asked one of my Christian friends to get me a copy of the New Testament, there was much hesitancy on his part. He refused at first and required a lot of pushing and pleading before he did it. When I got it, I had to hide it. Finally when I could read it, I had to do so in utter secrecy. This is why when it became known that I had embraced the Christian faith; I was arrested and went through a considerably hard time, finally being forced into exile by being deported from my country of birth. Sadly, even in the Western world comparatively few Christians are engaged in evangelizing Muslim friends. This may be due to fear or lack

of knowledge regarding how to go about it. And when they do, they are at a loss because the lines of distinction are blurred with seeming similarities. Yet the Christian faith and Islam are as far away from each other as the East is from the West." [178]

So you might be saying: "Why don't moderate Muslims speak up?" This question too, is a slippery one for followers of Islam. "Some years ago many leading Islamic scholars began to call upon Muslims to subject their faith to historical and critical scrutiny. Mounds of documents and texts were uncovered at that time, which compelled several scholars to begin a critical study of the very sources of the Koran. Notable scholars such as Ali Dashti from Iran, Nasr Abu Zaid, Egyptian professor of Arabic; Pakistani scholar Fazlur Rahman; Egyptian journalist Farag Foda; Algerian professor of law at the University of Paris Mohammed Arkoun; and Egyptian government minister and university professor Taha Hussein voiced some honest concerns about the texts and their meaning. These devout Muslim men have paid dearly for questioning the authenticity of the primary sources. Ali Dashti mysteriously disappeared in Iran. Abu Zaid was branded apostate and forced to flee the country with his wife (she would not have been permitted to remain with him once he was branded apostate). Farag Foda was assassinated, and so runs the list of those silenced by death or fear. When freedoms are so restricted in Islamic countries that no Muslim is allowed to disbelieve in Islam with impunity, is that a good Muslim law or a bad one? The freedom to believe – or not – is one of he most sacred privileges of the human mind. That freedom is taken away in the name of Islam. There is no religious freedom in most Islamic countries. Statistics that indicate that Islam is growing are not an accurate reflection of the hearts of the people, because they really have no choice. I had a close friend who was murdered because he became a Christian. Is this not just plain morally wrong? I well remember his sobering words to me shortly before he died: 'Brother Ravi, the more I see of the religion of the land of my birth, the more beautiful Jesus Christ looks to me.' If that does not provoke the extremists in Islam to take stock, I do not know what will."[182]

Similarly, my friend and Pakistani pastor Mujahid El Masih, who I recently worked with on a Voice of the Martyrs project, commented: *"Jihad is the great commission of Islam. Every Muslim is called to practice jihad. Jihad will never stop until there are no more non-Muslims in the world. Jihad is the vision of Islam and a major purpose of all Muslims on Earth. Jihad is the glue that brings Muslims together in defending Islam and expanding Islam."*[183] Mujahid was born in a nominally Christian home, became Muslim, joined the Pakistan military and then rediscovered Christ and became a pastor and attained his doctorate; he has also seen his church and Christian homes burned to the ground by fascist Islam in 1997, but this has only opened up his heart more as he goes around the United States speaking on Islam and Christianity. (Please visit his site www.ForTheLoveOfMuslims.com for more information)

While much more could be said on this topic, I will leave it at that and encourage the reader to research for themselves, but I think we can begin to see that Ravi's above quote is not without warrant when looking into what the 22nd Century may hold for Christianity as well as the rest of the world. We are seeing more Islamic countries developing nuclear weaponry, and some such as the president of Iran saying that to use these weapons to destroy all of Israel would be what Allah wants. While most of Iran wants to be rid of this oppressive regime of forced Islamic dictatorship, they can not achieve freedom of inquiry, religion, or even human rights on their own. Christ lived in a world of bitter turmoil much more than we are even in at the present, yet He never encouraged any type of physical violence, and therefore it is of my opinion that as Christians, we too can protest and make changes loudly and boldly without the "sword of steel" but with the "Sword of the Spirit/the Word of God;" but first, we must look to the past to help us understand the present, which in turn will help us to understand our possible future for tomorrow as well as into the 22nd century.

Most people (including Christians), do not realize that most of what we call the Middle East or the Islamic world was once Christian. The Christian faith originated in this part of the world often called "the East," and to this day, still has an often forgotten Christian presence alive in it. Phillip Jenkins has recently taken another look at this Christian east in his book "The Lost History of Christianity – the thousand year old golden age of the church in the Middle East, Africa, and Asia – and how it died." He describes how "when we move our focus away from Europe, everything we think we know about Christianity shifts kaleidoscopically, even alarmingly.[2]" Have we ever thought about this? The "church" was not born in Italy, Europe, or America; it was born in the Middle East. I know in our day and age Christianity in the Middle East seems almost like a contradiction in terms, but through the Middle Ages, a case could be made, that Iraq and the Middle East was just as much (if not more), a Christian cultural heartland, as was any specific part of Europe.

Unfortunately, as is all too common in the history of Christianity, inner turmoil and disputes (the polar opposite that Christ would endorse), would lead to a rift that would have enormous repercussions on Christianity... In the year of 325 A.D., the council of Nicaea reasserted the divinity of Christ. While no finite being truly understands the mystery of the infinite nor the Trinity associated with Father, Son, and Holy Spirit; many who had differing views on this subject were labeled "heretics." In 431 A.D. the Nestorians were cast out because of their differences. In 451 A.D. the council at Chalcedon defeated the Monophysites (Coptics and Syrians), and declared them too, "heretical." Much more division would take place all the way to the present, where everyone labels themselves and their organized church, "the church" and labels everyone else as "heretics…" While Christ preached "unity" His followers preached "division…"

"Almost from the beginning, persecution was part of the Christian experience. Nero, the Roman Emperor from AD 54-68, blamed Christians for the burning of Rome and had them imprisoned, thrown to wild animals, burned alive, crucified. Peter and Paul were martyred. Domitian, emperor from AD 81-96, persecuted Christians for refusing to worship him as divine. The Apostle John was exiled to Patmos. Emperor Marcus Aurelius (AD 161-80) refused to protect Christians from civil uprisings. Decius (AD 249-51) ordered thousands killed, including the Bishop of Rome, for not offering sacrifices to him as emperor. Diocletian (AD 284-305) tried to eliminate Christianity by ordering church buildings burned, Bibles confiscated, pastors tortured, and Christian civil servants stripped of citizenship. Those who refused to recant were executed. Tertullian, a second-century North African church leader and historian, grimily noted: 'Kill us, torture us, condemn us, grind us to dust… The more you mow us down, the more we grow.'" [139,140]

We are called upon to remember Hebrews 13:3 throughout this chapter: "Remember them that are in bonds, as bound with them; them that are ill-treated, as being yourselves also in the body." It is my hope that this first chapter will cause us to do just that.

Christianity has always been a "Western" religion, right?

"The Lord challenges us to suffer persecutions and to confess him. He wants those who belong to him to be brave and fearless. He himself shows how weakness of the flesh is overcome by courage of the Spirit. This is the testimony of the apostles and in particular of the representative, administrating Spirit. A Christian is fearless."

-Tertullian, (ca. 160 – ca. 220 A.D.)-

Once again, this book can do no justice in attempting to give even a concise history of Christianity or its theology, therefore I will attempt neither. Instead I will touch on a part of Christianity that is often overlooked. The "death of Christianity," in certain geographical areas of history, is not a subject often explored, nor unfortunately is it one of much interest to today's Christian or non-Christian alike, but it is a fact nonetheless. Many think of Christianity as a "Western Religion," but this only shows their ignorance of the historicity of Christianity, and its demographics as a whole. Christianity itself began and spread throughout the lands of Israel, Syria, Egypt, Asia Minor, and beyond. Soon "human" flaw slipped into the picture, and we find Christian attacking Christian. To summarize: the church fought against the church, (much like today, (minus the swords)), and as divisions were being made following the Council of Nicaea in 325 AD, two large Christian folds, the Nestorians and Monophysites (later called Jacobites), were cast out and deemed "heretical" in 431 and 451 AD as previously stated. We now had the beginnings of the Western vs. Eastern Christian Church; which leads ultimately to pushing Christ out of the picture.

Christianity in the 22nd Century

"When Westerners think about Christian history, most of them follow the Book of Acts in concentrating on the church's expansion west, through Greece and the Mediterranean world, and on to Rome. But while some early Christians were indeed moving west, many other believers – probably in greater numbers – journeyed east along the land routes, through modern-day Iraq and Iran, where they built great and enduring churches. Because of its location – close to the Roman frontier but just far enough beyond it to avoid heavy-handed interference – Mesopotamia or Iraq retained a powerful Christian culture at least through the 13th century. In terms of the number and splendor of its churches and monasteries, its vast scholarship and dazzling spirituality, Iraq was, through the late Middle Ages at least, as much a cultural and spiritual heartland of Christianity as was France or Germany or indeed Ireland. Iraq and Syria were the bases for two great transnational churches deemed heretical by the Catholic and Orthodox – namely, the Nestorians and Jacobites. And yet this older Christian world perished, destroyed so comprehensively that its memory is forgotten by all except academic specialists. During the Middle Ages, and especially during the 14th century, church hierarchies were destroyed, priests and monks were killed, enslaved, or expelled, and monasteries and cathedrals fell silent. As church institutions fell, so Christian communities shrank, the result of persecution and ethnic and religious cleansing. Survivors found it all but impossible to practice their faith..." [141]

So what was wrong with these Nestorians and other eastern groups? Were they really deserving of the title "heretic" and being cast out by the West? "The great Nestorian patriarch Timothy listed the fundamental doctrines that were shared by all the different groups – Nestorian, Monophysite, and Orthodox: all shared a faith in the Trinity, the Incarnation, baptism, adoration of the Cross, the holy Eucharist, the two Testaments; all believed in the resurrection of the dead, eternal life, the return of Christ in glory, and the last judgment. – We must never think of these churches as fringe sects rather than the Christian mainstream in large portions of the world. Among African and Asian Christians, these two strands of Christianity would certainly have outnumbered the Orthodox."[3]

This being said, we must remind ourselves that Christianity is not a "north or south or east or west" religion; but it is a "north and south and east and west" relationship with God. Many are the martyrs who have paved the way for us to know of Christ today, and to learn from them for the future of tomorrow. May we never forget that they are as much Christian as any one of us are today; their descendants still risk their lives every day for simply following Christ. The Christian churches were rapidly growing in the east throughout the continent of Asia, stopped only by internal conflict up until the mid 600s.

> Christianity has its origin in the Middle East and was the major religion of the region from the time of Jesus and for some time after the Muslim Conquests of the seventh century. Although Greek was the dominant language of the Early Church emerging from Hellenized communities around the Eastern Mediterranean (Anatolia, the Levant and Egypt), many Christian groups used other, local

languages, of which Syriac, Armenian, Coptic, Ge'ez, Georgian and Arabic are prominent.[4]

So why do many westerners and non-westerners alike claim that "Christianity is a western religion?" As I mentioned earlier (and will continue to do), it is simply due to an ignorance of Christianity's history and demographics, so I would highly encourage you to not simply take my word on this, but to research and educate yourself, your children, and your church on this very important topic. While we will touch on the subject of other religions briefly in chapter 2 of this book, it is important to look at history, to establish what are the largest threats we all face in truly following Christ as we approach the 22nd century.

Atheistic extremism, (in the forms of Nazism, communism, etc.), fascist Islam, and Christianity itself pose the biggest threats to Christians and Christianity in the past, present, and future as we will see. While Nazism and communism alone have claimed the lives of over 30,000,000 Christians in the last century, Islam has claimed a total of more than 9,000,000 Christians, (with some estimating over 200,000,000 total lives in the last 1,400 years), and on a solemn note, fellow Christians have taken the lives of over 5 million other Christians for differences of opinion. All three of these points should open our eyes, and cause us to fall on our knees in repentance and action. But before we address actions that we can take, we should review and pay tribute to those forgotten Christians that once thrived throughout Asia and the Middle East, and look at our past, to better understand our present, and ultimately, our future.

While this is not considered a politically correct book, I will call it a factual book, (that is why I have continually challenged the reader to review anything I am stating here); so let us be blunt:

"No religion poses a more formidable challenge to Christianity as we enter the third millennium than Islam. It is the dominant religion in many parts of the Third World, and its appeal as the religion of the oppressed, the religion for those who resent their treatment at the hands of the white European and American establishments, has considerable force. Christians need to support efforts to bring peace between Muslims and Jews in the Middle East and to foster understanding and acceptance between themselves and Muslims. Christian efforts to evangelize Muslims have so far had only marginal success and this might not change unless and until the climate of mutual suspicion between people of these two largest world religions changes.

Finally, Christians need to continue efforts to evangelize Muslims with the good news of Jesus Christ. We can and should work toward mutual understanding and acceptance while at the same time taking every opportunity to present the gospel to Muslims. This will require gaining a fair and accurate understanding of Islam as well as an ability to explain and defend the Christian truth claims over against the errors of Islam. Surely this largest of mission fields deserves the greatest of efforts and commitments of the Christian church as we enter the third millennium. [5]"

With that being said, let's review the historical challenges of Christianity's past leading up to the present.

The Founding of Islam and it's impact on Christianity

"I am sending you out like sheep among wolves. Therefore be as shrewd as snakes and as innocent as doves."

Matthew 10:16

While Christians were violently persecuted throughout their first two centuries of existence, most of their conflicts from 300-600 AD, (after Constantine's adoption of the Christian religion, thus making it legal); were internal strife and conflict with Christians fighting one another over doctrine and dogma, instead of keeping the focus on Christ. This carried on until the rise of a military-based religion known today as Islam. While we must remember that Muslims are not and never will be our enemies, it is important to address the 3 factors contributing and negating most to Christianity's history; as well as those that have contributed most to their martyrdom – starting with Christianity's history and that of Islam, with atheistic powers such as communism, and finally the internal strife of Christians against Christians. As mentioned earlier, these three groups have been the most responsible for Christian persecution; with the most tragic (in my opinion) being Christian against Christian. While the other groups did not claim to follow the peaceful precepts of the Christ, Christians did, so if you have read Foxe's Book of Martyrs, you will hear first-hand accounts of some of the most gruesome tortures of Christians, by other Christians. But we must start with the biggest crossroads that is still being played out today; and that is the crossroads of Christian history in the Middle East with the advent of Islam.

"In what came to be known as his farewell address, Muhammad is said to have told his followers: 'I was ordered to fight all men until they say there is no god but Allah.' This is entirely consistent with the Qur'an (9:5): 'Slay the idolaters wherever ye find them, and take them captive, and besiege them, and prepare for them each ambush.' In this spirit, Muhammad's heirs set out to conquer the world. In 570, when Muhammad was born, Christendom stretched from the Middle East all along North Africa, and embraced much of Europe. But only eighty years after Muhammad's death in 632, a new Muslim empire had displaced Christians from most of the Middle East, all of North Africa, Cyprus, and most of Spain. In another century Sicily, Sardinia, Corsica, Crete, and southern Italy also came under Muslim rule. How was this accomplished? How were the conquered societies ruled? What happened to the millions of Christians and Jews? [168]

As historian Thomas Madden states: "Unlike Islam, Christianity had no well-defined concept of holy war before the Middle Ages. Christ had no armies at his disposal, nor did his early followers. Only in AD 312, after the conversion to Christianity of the Roman Emperor Constantine I (for better or for worse), did the religion come into direct contact with statecraft and warfare. Within a century, Christianity and the Roman Empire were fused tightly together. Christians in government found themselves faced with questions of life and death, war and peace, questions that their religion had not wrestled with before. In the fifth century, St. Augustine outlined the necessary conditions for a Christian leader to wage a just war, (if there is a such thing), but he was quick to insist that the faithful not engage in wars of religious conversion or for the purpose of destroying heresies or pagans. Warfare was a necessary evil sometimes forced upon a good leader – it was not to be a tool of the church." Unfortunately, the church would ignore the outlook on war for most of its history. "The collapse of Roman power in the western Mediterranean did not spell a similar loss of power for Christianity because the Germanic barbarians who carved up the western empire were themselves Christians. The Christian faith, therefore, remained the faith of the large majority of people throughout Europe, the Near East, and North Africa. Christianity's first serious competitor did not arise until the seventh century, when the prosperous Arab merchant Mohammed founded a new religion.[6]"

Brigitte Gabriel (www.ActForAmerica.org), was a Christian from Lebanon that was driven out of Lebanon by an insurgence of fascist Islam while she was a teenager; she comments: "Twelve years after he received his prophethood, Mohammed decided to relocate to the city of Yathrib in the hope of recruiting more followers – especially the successful Jews and some Christians – to his religion. In 622 AD Mohammed emigrated with about fifty of his men to Yathrib from Mecca, a 200 – mile journey. This journey is referred to as the Hijra, which historians see as marking the beginning of an era in which Islam was spread not only as a religion but also as a political and military movement. The Hijra marked a turning point: the political reign of Islam had begun. It was at this point that Mohammed and his men began to fight and win battles to spread Islam and begin 1,400 years of Islamic terrorism and domination. Following the Hijra to Yathrib (which was later renamed Medina), which was then the center of Arab Jewish life, Mohammed aggressively tried to persuade the Jews to accept him as a true prophet and Islam as the true religion. In his attempt to win them over he adopted many of their customs and rituals, such as fasting, prohibiting the eating of pork, and circumcision.[7]"

Brigitte Gabriel, my wife and I in Denver, Colorado, (Sept 2009) **www.ActForAmerica.org**

Many people do not realize it, but since many of the principles and practices of Islam are derived from Syrian Christianity, Islam was largely tolerant to Christianity for quite some time. Other non-Abrahamic religions did not fare as well, however. The Zoroastrian religion existed throughout Persia at this time, and it saw its temples increasingly destroyed and replaced with mosques. Muslim regimes were also much harsher toward Manicheans, whom they hunted and killed.[8] But overall Islam was tolerant to those who wished to practice Christianity as long as they paid the "non-Muslim" taxes: "Non-Muslims after all survived as conquered tributary peoples, paying special taxes, so that it literally paid to keep these people in their infidel state.[9]"

This sadly was bound to change for a number of reasons, and we must remember that while much of Islam was tolerant to Christianity early on, this was not always the case. "After Mohammed's death in 632, Islam continued to grow, led by four caliphs. In 634 the second caliph Omar began referring to himself 'commander of the faithful.' With the sword and the Koran in hand, caliph Omar began a vicious expansion of Islam, conquering vast masses of land stretching from the Middle East into Africa.[10]" The battle of Yarmuck in 636, which gave the Muslims control of Syria, was a complete massacre, costing the lives of approximately 50,000 soldiers of the Christian Byzantine Empire.[11]

All this time, Christians were still focusing on internal strife and who was the "heir of Christendom," which further alienated them from one another. We then see that by the year 637 Muslims entered the city of Jerusalem (in which the Dome of the Rock was completed in 692), 642 Muslims capture Alexandria, 647 Muslims invaded North Africa, 651 Muslims completed the conquest of Persia; we then see the bloody Umayyad Caliphate established in Damascus in 661. Muslims then turn to siege Constantinople from 647-77 but fail; they take Carthage in 698, conquered Spain in 711, and established but are repulsed again in 717 as they yet again lay siege to Constantinople. Many Christians and non-Christians do not realize that the Muslim armies were within 100 miles of Paris at this point in history, when the Frankish general Charles Martel was able to defeat the Muslims at the Battle of Tours in 732, which turned the tide of the conquest of Europe, and proceeded to drive Islamic forces back out of the lands they had invaded.

> *"Muslim conquerors who swept through all of Christian North Africa also crossed the straits of Gibraltar and established their rule over Spain. By the eighth century, Muslim expeditionary forces were crossing the Pyrenees and marching into the heart of Catholic Europe. In 732, at the famous Battle of Tours, Frankish leader Charles Martel defeated the Muslims, driving them back to Spain.*[12]*"*

Who will be the next Charles Martel?

Charles Martel (ca. 688 – 22 October 741), called Charles the Hammer, was a Frankish military and political leader who served as Mayor of the Palace under the Merovingian kings and ruled *de facto* during an interregnum (737–43) at the end of his life, using the title Duke and Prince of the Franks. In 739 he was offered the title of Consul by the Pope, but he refused.[13] He is perhaps best remembered for winning the Battle of Tours in 732, in which he defeated an invading Muslim army and halted northward Islamic expansion in western Europe.[14]

When I say "Who will be the next Charles Martel," I do mean this in a Christ-like approach; with the sword of truth and spirit, not the sword of steel and death. I pray that somewhere, (perhaps China in their Back to Jerusalem movement), that a "Billy Graham, Dietrich Bonhoeffer, Samuel Zwemer, or Martin Luther King Jr.," will rise up, and similar to Charles Martel's stopping of fascist Islam's conquering all of Europe, will help to tear down the walls of fascist Islamic oppression that has enveloped the whole of the Middle East and is leaking increasingly into Europe and beyond but through love, peace, and most importantly, with the "truth" of Christ, Christians need not "force" any religion dogmas on anyone but allow them the choice. When my Muslim friend recently read where Jesus told those who were wanting to stone a woman caught in adultery that "whoever is without sin, may cast the first stone," she was moved with total compassion. Even now when she begins to defend something Islamic happening in the news and I ask her "why not take Jesus' approach to it," she stops, sighs, and says, "yeah... that is true." So if history repeats, we desperately need a loving type of "Charles Martel" bearing Christ's message in the Islamic lands of Iran, Afghanistan, Yemen, and beyond, to show the "truth" and that the Truth will "set them free." Should these Islamic lands not at least have the chance to choose for themselves who they will follow?

Historian Paul K. Davis said concerning Martel, in *100 Decisive Battles* "Having defeated Eudes, he turned to the Rhine to strengthen his northeastern borders - but in 725 was diverted south with the activity of the Muslims in Acquitane." Martel then concentrated his attention to the Umayyads, virtually for the remainder of his life.[15] Indeed, 12 years later, when he had thrice rescued Gaul from Islamic Umayyad invasions, Antonio Santosuosso noted when he destroyed an Umayyad army sent to reinforce the invasion forces of the 735 campaigns, "Charles Martel again came to the rescue".[16] It has been noted that Charles Martel could have pursued the wars against the Saxons—but he was determined to prepare for what he thought was a greater danger, in that of Islam.

It is important to note, that while war should always be a last option for the Christian, many historians, including Sir Edward Creasy, believe that had he failed at Tours, Islam would probably have overrun Gaul, and perhaps the remainder of Western Europe. Gibbon made clear his belief that the Umayyad armies would have conquered from Rome to the Rhine, and even England, having the English Channel for protection, with ease, had Martel not prevailed. Creasy said "the great victory won by Charles Martel ... gave a decisive check to the career of Arab conquest in Western Europe, rescued Christendom from Islam." Gibbon's belief that the fate of Christianity hinged on this battle is echoed by other historians including John B. Bury, and was very popular for most of modern historiography. It fell somewhat out of style in the twentieth century, when historians such as Bernard Lewis contended that Arabs had little intention of occupying northern France. More recently, however, many historians have tended once again to view the Battle of Tours as a very significant event in the history of Europe and Christianity. Equally, many, such as William Watson, still believe this battle was one of macro historical world-changing importance, if they do not go so far as Gibbon does rhetorically.

> "A Muslim France? Historically, it nearly happened. But as a result of Martel's fierce opposition, which ended Muslim advances and set the stage for centuries of war thereafter, Islam moved no farther into Europe. European schoolchildren learn about the Battle of Tours in much the same way that American students learn about Valley Forge and Gettysburg."[18]

In the subsequent decade, Charles led the Frankish army against the eastern duchies, Bavaria and Alemannia, and the southern duchies, Aquitaine and Provence. He dealt with the ongoing conflict with the Frisians and Saxons to his northeast with some success, but full conquest of the Saxons and their incorporation into the Frankish empire would wait for his grandson Charlemagne, primarily because Martel concentrated the bulk of his efforts against Muslim expansion.[19]

After being pushed back out of most of Europe, Muslim forces then turned their eyes to China where they defeated the Chinese at the Battle of Talas River in 751, and proceeded into parts of India and beyond. While many of these Christian histories were preserved, (at least in part), there were examples of such great cities as Carthage, (where church father Tertullian resided among others), which served as a center for Christian thought and influence, that following the Muslim conquest in 698 virtually disappeared completely from the history.[20] How sad is it that none of these stories are repeated in Western churches today, either through ignorance or indifference?

With Christianity having to fight Muslim forces on every corner, they sadly chose to continue to fight among each other as well during this pivotal point in history; as well as follow their own interests instead of following Christ. An example of this internal strife in Christendom can be heard in the expression of Michael the Syrian (a Jacobite) when he reported:

> "The God of vengeance... seeing the wickedness of the Romans who, wherever they ruled, cruelly robbed our churches and monasteries and condemned us without pity, raised from the region of the south the children of Ishmael to deliver us by them from the hands of the Romans... it was no light advantage for us to be delivered from the cruelty of the Romans."[21]

This sad internal strife was realized to its fullness when in 1054 the two kingdoms of the West split into two kingdoms, (further splitting themselves from the Christian East), in what came to be known as the East-West Schism. (Are we bound to repeat a similar mistake today?)

Christianity's Split and its Impact on the past/ present/future of Christians

Are we seeing a pattern yet? We have Christian Brothers/ Sisters dying throughout the world, with an average of 150,000 Christian being martyred every year in Islamic lands, while in the West/United States, we are more worried about arguing over what praise instruments a church should or should not use. Is this truly what Christ would want? Yet again, let's see what internal strife existed among each other while outside forces of fascism were pressing us on all sides, and resulted when we repeated a similar example in the not so distant past; that we may better understand its impact (again) on our future. The East–West Schism[22] divided medieval Christianity into Eastern (Greek) and Western (Latin) branches, which later became known as the Eastern Orthodox Church and the Roman Catholic Church, respectively. Relations between East and West had long been embittered by political and ecclesiastical differences and theological disputes. Pope Leo IX and Patriarch of Constantinople Michael Cerularius heightened the conflict by suppressing Greek and Latin in their respective domains. In 1054, Roman legates travelled to Cerularius to deny him the title Ecumenical Patriarch and to insist that he recognize the Church of Rome's claim to be the head and mother of the churches. Cerularius refused. The leader of the Latin contingent, Cardinal Humbert, excommunicated Cerularius, while Cerularius in return excommunicated Cardinal Humbert and other legates. The Western legates' acts might have been of doubtful validity due to Leo's death, while Cerularius's excommunication applied only to the legates personally.[23] Still, the Church split along doctrinal, theological, linguistic, political, and geographical lines, and the fundamental breach has never been healed.[24]

We now begin to see a "divide, conquer, and destroy" approach being carried out through the Christian East. (Please note, this was not merely militarily, but also educationally; which is still seen today, where 52 countries are either hostile or outlaw Christianity's practices. See Appendices)

Islam sees a real threat of all the facts being placed on the table, because philosophy, history, and evidences were on the side of the Christian, so the best way to handle this is to keep the people ignorant and close all debate. (This is carried through to today where under Sharia (Muslim) law, no churches, no freedom of religion, no open dialogue to educate the people to both sides is allowed). Muslim activist al-Jahiz complained that Christians: "hunt down what is contradictory in our traditions, our reports with a suspect line of transmission and the ambiguous verses of our scripture. Then they single out the weak-minded among us and question our common people concerning these things... and they will often address themselves to the learned and powerful among us, causing dissension among the mighty and confusing the weak."[25]

Since Islam is a direct off-shoot of Judaism and Christianity, (similar to Mormonism), it retained at least moderate relations with Christianity as mentioned earlier, up until this point when it began to systematically convert by either policies or the sword. This however was not exclusively attacks on religion; in several instances the hostilities were merely increased by the competition of Christianity in the bordering Byzantine Empire in Asia Minor and Northern Africa. Seljuk Turks began raiding and warring with much of the empire which happened to be Christian. This proved devastating to Christian society.[26]

Common sense should tell us that with the vast majority of the Middle East, modern day Turkey, and Northern Africa being Christian then, and Muslim now, was not a mere accident. Did all the Christians simply decide to move away, and

Islam moved in – Of course not. My prayer is that we not only remember these martyrs, but learn from them, and from the plight of the persecuted church today that so many either neglect, or are simply ignorant of. The persecution intensified at this point.

> *"Whole Christian communities were annihilated across central Asia, and surviving communities shrank to tiny fractions of their former size. Christianity now disappears in Persia, and across southern and central Iraq.[27] (Around the year 1340)"*

One might ask, "how did Asia Minor (modern Turkey) go from being mostly Christian to today being 99% Islamic?" Let's quickly look at the town of Ephesus, where Paul preached to see what happened to many of its Christian inhabitants:

> *"In 1304, Turkish forces obliterated the city of Ephesus, where Paul once confronted a mob chanting the glories of Artemis: ALL Christians were either killed or deported.[28]"*

At this point in history the Islamic warlord known as Timur or Tamerlane, would prove to be the Hitler or Stalin of his day, and crush the remnants of Christianity to an unprecedented degree, claiming over five million Christian Martyrs from 1370-1405. The final nail in the coffin came when Christians (still fighting amongst themselves), would leave the last remnants of the Christian East's capital Constantinople, in a state of weakness, having been sacked by their "Christian" Brothers to the West:

Speros Vryonis in *Byzantium and Europe* gives a vivid account of the sack of Constantinople by the Frankish and Venetian Crusaders of the Fourth Crusade:

The Latin soldiery subjected the greatest city in Europe to an indescribable sack. For three days they murdered, raped, looted and destroyed on a scale which even the ancient Vandals and Goths would have found unbelievable. Constantinople had become a veritable museum of ancient and Byzantine art, an emporium of such incredible wealth that the Latins were astounded at the riches they found. Though the Venetians had an appreciation for the art which they discovered (they were themselves semi-Byzantines) and saved much of it, the French and others destroyed indiscriminately, halting to refresh themselves with wine, violation of nuns, and murder of Orthodox clerics. The Crusaders vented their hatred for the Greeks most spectacularly in the desecration of the greatest Church in Christendom. They smashed the silver iconostasis, the icons and the holy books of Hagia Sophia, and seated upon the patriarchal throne a whore who sang coarse songs as they drank wine from the Church's holy vessels. The estrangement of East and West, which had proceeded over the centuries, culminated in the horrible massacre that accompanied the conquest of Constantinople. The Greeks were convinced that even the Turks, had they taken the city, would not have been as cruel as the Latin Christians.
The defeat of Byzantium, already in a state of decline, accelerated political degeneration so that the Byzantines eventually became an easy prey to

the Turks. The Crusading movement thus resulted, ultimately, in the victory of Islam, a result which was of course the exact opposite of its original intention.[29]

How anyone who follows Christ could think this infighting could be in anyway what Christ would want, is mind boggling. The last remnants of the Byzantium Christian Empire had now been reduced to the remains of the city of Constantinople, which was a shell of its former glory, and sacking by fellow Christians' hands. The Eastern Emperors appealed to the west for help, but the Pope would only consider sending aid in return for a reunion of the Eastern Orthodox Church with the See of Rome. Church unity was considered, and occasionally accomplished by imperial decree, but the Orthodox citizenry and clergy intensely resented the authority of Rome and the Latin Rite. Some western troops arrived to bolster the Christian defense of Constantinople, but most Western rulers, distracted by their own affairs, did nothing as the Ottomans picked apart the remaining Byzantine territories.

Constantinople by this stage was under populated and dilapidated. The population of the city had collapsed so severely that it was now little more than a cluster of villages separated by fields. On 2 April 1453, Sultan Mehmed's army of some 80,000 men and large numbers of irregulars laid siege to the city. Despite a desperate last-ditch defense of the city by the massively outnumbered Christian forces (c. 7,000 men, 2,000 of whom were foreign), Constantinople finally fell to the Ottomans after a two-month siege on 29 May 1453.[30] The last Byzantine emperor, Constantine XI Palaiologos, was last seen casting off his imperial regalia and throwing himself into hand-to-hand combat after the walls of the city were taken.[31]

> *"The Byzantines were human like the rest of us, victims of the same weaknesses and subject to the same temptations, deserving of praise and of blame much as we are ourselves. What they do not deserve is the obscurity to which for centuries we have condemned them. Their follies were many, as were their sins; but much should surely be forgiven for the beauty they left behind them and the heroism with which they and their last brave Emperor met their end, in one of those glorious epics of world history that has passed into legend and is remembered with equal pride by victors and vanquished alike. That is why five and a half centuries later, throughout the Greek world, Tuesday is still believed to be the unluckiest day of the week; why the Turkish flag still depicts not a crescent but a waning moon, reminding us that the moon was in its last quarter when Constantinople finally fell; and why, excepting only the Great Church of St. Sophia itself, it is the Land Walls – broken, battered, but still marching from sea to sea – that stand as the city's grandest and most tragic moment."* [179]

Hopefully, we can all see that seeking internal strife and conflict amongst each other when fascist Islam or atheistic communism is knocking on our door - did not benefit Christianity at all in the past, nor will it in our future, (nor is it an example of following Christ (from whom Christ-ianity bears its name)).

At this point Christian Europe felt it had to answer the political and military threat of the Islamic empire. The Islamic invasions of Spain lasted more than 900 years (710-1616 AD), bringing Muslim domination and enslavement of Christians and Jews. Almost as many Christians and Jews were killed by Muslims in only one day in Granada as all the Spaniards killed (5,000) during the more well-known Spanish Inquisition.[32] Following Martel's victory at Tours, Europe concentrated its efforts on taking control of Jerusalem and on driving the Moors from Spain. Jerusalem changed hands a couple of times and ended up in control by the Muslims, where it remained (under different ethnic powers) from the thirteenth to the twentieth century. The Europeans were more successful in Spain, eventually expelling the Moors in 1492 (the same year as Columbus's first expedition).[33]

> *"I tell you the truth, unless a kernel of wheat falls to the ground and dies, it remains only a single seed. But if it dies, it produces many seeds. [25]The man who loves his life will lose it, while the man who hates his life in this world will keep it for eternal life.*
> **John 12:24-25**

Protestant Christian vs. Roman Catholic Christian vs. Eastern Orthodox Christian, Etc.

This leads up to the time of the Reformation, more fighting between Christian and Christian - Catholic and Protestant, and of course the founding of colonies in the Americas. From my research, I have heard of the most extreme measures of torture and brutality carried out by Christians against Christians. It becomes painfully obvious after reading John Foxe's Book of Martyrs how few who claimed to be Christians really followed Christ. Once again we see Christians fighting Christians resulting in thousands upon thousands of deaths. (I would highly suggest reading Foxe's famous Book of Martyrs, which gives account of some of the most disturbing and cruel butchery of Christian against Christian (Catholic against Protestant), to educate ourselves of what Christianity without Christ can become). While not nearly as many Christians were martyred by other Christian in sheer numbers than by atheists or Muslim forces, the tortures that were of the most gruesome nature that I have heard in all of my research, was that of Christian killing Christian.

A quick example of this is when the Roman Catholic Church moved to suppress the Cathar heresy, the Pope having sanctioned a crusade against the Albigensians; during the course of which the massacre of Beziers took place, with between seven and twenty thousand deaths. (This was the occasion when the papal legate, Arnaud Amalric, asked about how Catholics could be distinguished from Cathars once the city fell, famously replied, "Kill them all, God will know His own."). Over the twenty year period of this campaign an estimated 200,000 to 1,000,000 people were killed.[34]

In the 1572 St. Bartholomew's Day Massacre the French king ordered the murder of Protestants in France. Intolerance of dissident forms of Protestantism continued, as evidenced by the exodus of the Pilgrims who sought refuge in America, founding the Plymouth Colony in Massachusetts in 1620. In the modern period, such events include violence between Mormons and Protestants in the United States during the 19th century.[35]

While I will not further attempt to summarize these atrocities in this book that were carried out during this time period, it should serve as another solemn warning to us all the role that Christians against Christians has played in Christian history.

Following this time period of Christian persecuting Christian, it was the British Empire which would eventually contain the expansion of the Islamic powers worldwide. In 1600 the British established the East India Company, establishing economic interests in a part of the world which was increasingly under Muslim control. By the time of the American Revolution the Muslim Ottoman Empire was breaking up gradually, due in large part to British (and French) colonialism in India, Africa, and Southeast Asia. In 1858 India became part of the British empire, and between 1880 and 1918 the British gained control of both Egypt and Palestine. The Ottoman Empire disappeared completely after World War I, and during the 1920s several independent states emerged, typically with the British or French lending support. These included Turkey, established as a secular state, and several Arab monarchies with only partially Islamic systems of law (e.g., Saudi Arabia, Egypt, Iraq, Syria).

The British intent in taking control of Palestine all along was to open the way for Jews to establish a homeland there, an intention formally declared in the Balfour Declaration (1917). Already in 1881 Jewish immigrants had begun building new settlements in Palestine, still largely Arab Muslim in population. The Jewish population grew in Palestine after World War I and increased dramatically during the Nazi regime in the 1930s and throughout World

War II, as Jews sought to escape the Holocaust. After the war ended, in 1947, the United Nations partitioned Palestine into Jewish and Arab states, giving Jews 52 percent of the land, and in 1948, the nation of Israel began its modern existence. The Arab Palestinians rejected this plan and in 1948 Jordan, Syria, Lebanon, Egypt, and Iraq launched war against Israel. In less than a year Israel had won the war and enlarged its borders to encompass some 77 percent of the former Palestine. Meanwhile, in India the British-educated Mohandas K. ("Mahatma") Gandhi led a nonviolent resistance movement to pressure the British for Indian independence. Although the British did grant India its independence, Gandhi's hope that Indian Muslims and Hindus would live together in peace was not realized. The Muslim-dominated areas in the east and west wings of the Indian subcontinent became an independent Pakistan in 1947, and Gandhi was himself assassinated in 1948.[36]

*[I feel that I should make clear, that I am recording factual events; I do not want to even suggest that I am preaching any type of hate against any group, (Muslim, Communist, or even fellow Christian denominations). I am simply giving you a glimpse of what these 3 top persecutors of Christ's followers have been: Atheism, Islam, and Christianity. I simply ask us to all reflect on what this history meant then and now. Don't get me wrong, I do not support Islam, Communist or atheistic regimes, nor un-Christ-like Christian groups, and I pray that we will see a reform in Islam as well as these other groups, (see Appendix D) But I do want to clarify that if I say or quote: "Islam destroyed many cities," or "Christians butchered other Christians in Paris in 1600," this is a factually documented statement; and that while my purpose is not in any way to preach any form of hate nor judge anyone past or present; I do hope that both believer and skeptic will see the dangers in fascist Islam, atheism, and a non-authentic

Christian faith, and educate themselves about each. Christ at no time supported sin, but He always loved the sinner; and as Richard Wurmbrand said, "He hates Communism, but loves the Communist;" I too, hope that you pick up this type of sentiment in this book.]

At this point in history, please keep in mind that the Byzantine Empire is now destroyed, and the heart of Orthodox Christianity has been moved to Russia. We fast forward through the reformation period, to the late 1700's when the French Revolution is underway; this will be the seed of where we start the next chapter.

De-Christianization of France during the French Revolution leads to a new type of Christian persecution by atheistic governments

ETHICS by Dietrich Bonhoeffer pages 102-105:

"The French revolution created a new unity of mind in the west. This unity lies in the emancipation of man as reason, as the mass, as the nation. In the struggle for freedom these three are in agreement, but once their freedom is achieved they become deadly foes. Thus the new unity already bears within itself the seeds of decay. Furthermore, there becomes apparent in this an underlying law of history, namely that the demand for absolute liberty brings men to the depths of slavery. The master of the machine becomes its slave. The machine becomes the enemy of men. The creature turns against its creator in a strange re-enactment of the Fall. The emancipation of the masses leads to the reign of terror of the guillotine. The liberation of man as an absolute ideal leads only to man's self-destruction. At the end of the path which was first trodden in the French Revolution there is nihilism."

We have seen the effects of Islam and non-Christ Christianity in the form of fascism that began to stagger into the 16th century and was replaced temporarily by a new threat to not just Christianity, but to the world itself when left unchecked to carry its ideology to whatever end it chooses - atheism.

"By far the most direct and radical challenge to the Christian faith is to deny the existence of any God. In the twentieth century the atheistic worldview that rejects all beliefs in supernatural or transcendent beings achieved influence far greater than at any time previously in the West. Even in the twentieth century, though, its influence has far outstripped its adherence in terms of sheer numbers. While atheists remain in the minority in all Western countries, they have had an inordinate influence on the culture as the most forceful advocates of the secularization of society."[38]

What we see with the French revolution is a rebellion against the state, and many interpreted Christ with the state, (due to the un-authentic Christianity enforced by many European governments up to this point, (as mentioned earlier)). As I studied in my undergraduate class "The Era of the French Revolution," we see the seeds of France's, (and Europe's), moral collapse being traced back to the revolution in many ways; as Dietrich Bonhoeffer described in his book Ethics:

The new unity which the French Revolution brought to Europe – and what we are experiencing today is the crisis of this unity – is therefore western godlessness. It is totally different from the atheism of certain individuals (of the past); It is not the theoretical denial of the existence of a God. It is itself a religion, a religion of hostility to a God. It is in just this that it is western. In the form of all possible Christianities, whether they be nationalist, socialist, rationalist or mystical, it turns against the living God of the Bible, against Christ. Its God is the New Man, no matter whether he ears the trademark of Bolshevism or of Christianity.

With the destruction of the biblical faith in God and of all divine commands and ordinances, man destroys himself. There arises an unrestrained viatlism which involves the dissolution of all values and achieves its goal in final self-destruction, in the void." (pg 102-5)

So what happens when atheism has full freedom? You hear some individuals such as Bill Maher speaking from ignorance when they insinuate that a world without religion would be a more utopian state. We have had such governments in the past and we have seen that they are far from "utopian." *(It is quite alarming to see that a predominantly Christian nation such as France, Germany, or Russia can change so quickly, (a lesson we should all take note of)).*

The De-Christianization of France during the French Revolution is a conventional description of the results of a number of separate policies, conducted by various governments of France between the start of the French Revolution in 1789 and the Concordat of 1801, forming the basis of the later and less radical Laïcité movement. The goal of the campaign was the destruction of Catholic religious practice and of the religion itself.[39] The program included the following policies:

- the deportation of clergy and the condemnation of many of them to death,
- the closing, desecration and pillaging of churches, removal of the word "saint" from street names and other acts to banish Christian culture from the public sphere
- removal of statues, plates and other iconography from places of worship
- destruction of crosses, bells and other external signs of worship
- the institution of revolutionary and civic cults, including the Cult of Reason, and subsequently, the Cult of the Supreme Being,
- the large scale destruction of religious monuments,
- the outlawing of public and private worship and religious education,
- forced marriages of the clergy,

- forced abjurement of priesthood, and
- the enactment of a law on October 21, 1793 making all nonjuring priests and all persons who harbored them liable to death on sight.[40]

(You hear many people reference this time as the "Enlightenment," when atheism became a more viable option for many. It is important to note that today's militant atheist (as the past's) was made "viable" because of the church's abuse of power, the hypocrisy that plagued the church on every side, and simply not following Christ. Christianity is just a word, and if it does not follow its founder (Christ Jesus) then it is 100% without warrant. This "De-Christianization" that started predominantly in France, and its presence today, is due to past "Christians" not being "Christ-ians." May we learn very carefully from this lesson, and work to un-do what we have "messed up" in the first place.)

The climax of this French revolution from religion was reached with the celebration of the Goddess "Reason" in Notre Dame Cathedral on 10 November 1793. Under threat of death, imprisonment, military conscription or loss of income, about 20,000 constitutional priests were forced to abdicate or hand over their letters of ordination and 6,000 - 9,000 were coerced to marry, many ceasing their ministerial duties. Some of those who abdicated covertly ministered to the people. By the end of the decade, approximately 30,000 priests were forced to leave France, and thousands who did not leave were executed.[41] Most of France was left without the services of a priest, deprived of the sacraments and any nonjuring priest faced the guillotine or deportation to French Guiana.[42]

Darwinism/Marxism/Freudism helped to usher in the atheistic Russian revolution of 1917

After the Revolution of 1917, the Bolsheviks undertook a massive program to remove the influence of the Russian Orthodox Church from the government and Russian society, and to make the state atheist. Thousands of churches were destroyed or converted to other uses, and many members of clergy were imprisoned for anti-government activities. An extensive education and propaganda campaign was undertaken to convince people, especially the children and youth, to abandon religious beliefs. This persecution resulted in the martyrdom of millions of Orthodox followers in the 20th century by the Soviet Union, whether intentional or not. This persecution spread not only to the Orthodox, but also other groups, such as the Mennonites, who largely fled to the Americas.[43]

Before and after the October Revolution of November 7, 1917 (October 25 Old Calendar) there was a movement within the Soviet Union to unite all of the people of the world under Communist rule. This included the Eastern European bloc countries as well as the Balkan States. Since some of these Slavic states tied their ethnic heritage to their ethnic churches, both the peoples and their church were targeted by the Soviet and its form of State atheism. The Soviets' official religious stance was one of "religious freedom or tolerance," though the state established atheism as the only scientific truth.[44] Criticism of atheism was strictly forbidden and sometimes resulted in imprisonment.[45] Some of the more high profile individuals executed include Metropolitan Benjamin of Petrograd, Priest and scientist Pavel Florensky and Bishop Gorazd Pavlik.

The Soviet Union was the first state to have as an ideological objective the elimination of religion. Toward that end, the Communist regime confiscated church property, ridiculed religion, harassed believers, and propagated atheism in the schools. Actions toward particular religions, however, were determined by State interests, and most organized religions were never outlawed. It is estimated that 21 million Russian Orthodox Christians were martyred in the gulags by the Soviet government, not including torture or other Christian denominations killed.[46] Some actions against Orthodox priests and believers along with execution included torture, being sent to prison camps, labor camps, or mental hospitals.[47] The result of this militant atheism was to transform the Church into a persecuted and martyred Church. In the first five years after the Bolshevik revolution, 28 bishops and 1,200 priests were executed.[48]

As mentioned earlier, the main target of the anti-religious campaign in the 1920s and 1930s was the Russian Orthodox Church, which had the largest number of faithful. A very large segment of its clergy, and many of its believers, were shot or sent to labor camps. Theological schools were closed, and church publications were prohibited. In the period between 1927 and 1940, the number of Orthodox Churches in the Russian Republic fell from 29,584 to less than 500. Between 1917 and 1940, 130,000 Orthodox priests were arrested. The widespread persecution and internecine disputes within the church hierarchy lead to the seat of Patriarch of Moscow being vacant from 1925-1943.

After Nazi Germany's attack on the Soviet Union in 1941, Joseph Stalin revived the Russian Orthodox Church to intensify patriotic support for the war effort. By 1957 about 22,000 Russian Orthodox churches had become active. But in 1959 Nikita Khrushchev initiated his own campaign against the Russian Orthodox Church and forced the closure of about 12,000 churches. By 1985 fewer than 7,000 churches remained active.[48]

Hitler and the Nazis on the other hand, enjoyed widespread support from traditional Christian communities, mainly due to a common cause against the anti-religious German Bolsheviks. Once in power, the Nazis moved to consolidate their power over the German churches and bring them in line with Nazi ideals. Dissenting Christians such as Dietrich Bonhoeffer, went underground and formed the Confessing Church, which was persecuted as a subversive group by the Nazi government. Many of its leaders were arrested and sent to concentration camps, and left the underground mostly leaderless. Church members continued to engage in various forms of resistance, including hiding Jews during the Holocaust and various attempts, largely unsuccessful, to prod the Christian community to speak out on the part of the Jews.

The Catholic Church was particularly suppressed in Poland because of the Church's opposition to many of the Nazi Party's beliefs. Between 1939 and 1945, an estimated 3,000 members, 18% of the Polish clergy, were murdered; of these, 1,992 died in concentration camps.[49] In the annexed territory of *Reichsgau Wartheland* it was even harsher than elsewhere. Churches were systematically closed, and most priests were killed, imprisoned, or deported to the General Government. The Germans also closed seminaries and convents persecuting monks and nuns throughout Poland. In Pomerania, all but 20 of the 650 priests were shot or sent to concentration camps. Eighty percent of the Catholic clergy and five of the bishops of Warthegau were sent to concentration camps in 1939; in the city of Breslau (*Wrocław*), 49% of its Catholic priests were killed; in Chełmno, 48%. Protestants in Poland did not fare well either. In the Cieszyn region of Silesia every single Protestant clergy was arrested and deported to the death camps.[49] Not only were Polish Christians persecuted by the Nazis, in the Dachau concentration camp alone, 2,600 Catholic priests from 24 different countries were killed.[49]

Christianity in the 22nd Century

Hitler Speaks (1933) by Hermann Rauschning:

"I tell you: we must prevent the churches from doing anything but what they are doing now, that is, losing ground day by day. Do you really believe the masses will ever be Christian again? Nonsense! Never again. That tale is finished. No one will listen to it again." The church will be made to dig their own graves. They will betray their God to us. They will betray anything for the sake of their miserable little jobs and incomes."

I promise you, he concluded, *'if I wished to, I could destroy the church in a few years; it is hollow and rotten and false through and through. One push and the whole structure would collapse. I shall give them a few years reprieve. Why should I quarrel? They will swallow anything in order to keep their material advantages. Matters will never come to a head. They will recognize a firm will, and we need only show them once or twice who is the master. Then they will know which way the wind blows. They are no fools. The Church was something really big. Now we're its heirs. Its day is gone. AS LONG AS YOUTH FOLLOW ME, I DON'T MIND IF THE OLD PEOPLE LIMP TO THE CONFESSIONAL. BUT THE YOUNG ONES – THEY WILL BE DIFFERENT. I GUARANTEE THAT." –Adolf Hitler (pg 61)*

(Interesting to note that less than half of Germans even believe in God, and it is also the second largest European Muslim country behind France as of 2009)

China's Persecution of Christians

This form of atheistic government is found today in the People's Republic of China, where sadly, many Americans and our government turn a blind-eye to human rights atrocities and religious freedoms in the name of economic interests.

The communist government of the People's Republic of China tries to maintain tight control over all religions, so the only legal Christian Churches (Three-Self Patriotic Movement and Chinese Patriotic Catholic Association) are those under the Communist Party of China control. Churches which are not controlled by the government are shut down, and their members are imprisoned. In 2009, Christians must worship in registered, regulated churches. According to the Jubilee Campaign, an interdenominational lobby group, about 300 Christians caught attending unregistered churches were in jail in 2004.[50] Gong Shengliang, head of the South China Church, was sentenced to death in 2001. Although his sentence was commuted to a jail sentence, Amnesty International reports that he has been tortured.[50]

With the near destruction of Christianity and the advent of nihilism carried to its logical ends in Russia by the atheistic Soviet Union, and now adopted by China; I think it is fair to say that a world with no focus on God, is anything but a "utopia."

However, we are seeing another story in how the Lord is using China to spread His message and possibly lead His church in the future. What we should find somewhat dumfounding at this point, is that persecution does indeed appear to strengthen the true Christian. We find in China a mere population of about 700,000 Christians by 1950, and then after the atheistic Chinese regime vehemently persecuted the church for half a century, the population is approximately 100,000,000 Christians, while as Hitler mentioned above;

we tragically see Germany with not much over 1/3 who even believe in God; they completely lost their identity in many ways, it would seem. So now we begin to see why Christ said to rejoice if you are persecuted for His name.

Many Chinese have concluded that their part of Christian history will be to take the Gospel throughout all of Asia and the Middle East. They feel they are in a time in history where they may succeed; in sharing the Gospel message with the Middle East on their way to Jerusalem. This movement has now been coined "The Back to Jerusalem movement." One of the Chinese believers commented that while the west may know evangelism, they know suffering and persecution, and will thus be successful in their outreach to the Islamic nations. Let us pray that they succeed, and that we support them with prayer, and every means that we can; and moreover, that we may learn from them – learn from their example to sacrifice all for Christ, (regardless of our materials, church buildings, retirement, etc). The book "Back to Jerusalem" describes it this way:

> "It seems to be an unfortunate characteristic of many Christians that when things are going well we like to stop, make ourselves comfortable, and enjoy our successes. The gospel had saturated Jerusalem, but the disciples were starting to forget the other stages of the Great Commission. To help them remember, the Lord provided some persecution! On the same day that Stephen became the first martyr of the church, 'a great persecution broke out against the church at Jerusalem, and all except the apostles were scattered throughout Judea and Samaria.'"[138]

Pastor Richard Wurmbrand reiterates this point reflecting on his 14 years in communist prisons:

> *"Men with such ordination, who have never had any theological training and who very often know little of the Bible, carry on the work of Christ. It is like the Church in the first centuries. What seminaries did those who turned the world upside-down for Christ attend? Did they all know how to read? And from where did they receive Bibles? God spoke to them.[51]"* (Tortured for Christ, pg 94)

Praise the Lord that this year marks the 20th anniversary of the fall of the atheistic state of Russia; but let us not forget it also marks the 20th anniversary of the Tiananmen Square massacre. May we not forget our brave Brothers and Sisters in China suffering under an atheistic government today. Perhaps the persecuted Chinese can teach us more than we can teach them from the safety and security of our church pews?

Two Chinese believers who have both spent time in prison, summed up this philosophy well in the Back to Jerusalem movement:

> *"We're totally committed to planting groups of local believers who meet in homes. We have no desire to build a single church building anywhere! This allows the gospel to spread rapidly, is harder for the authorities to detect, and enables us to channel all our resources directly into gospel ministry."* (pg 58)

> "We have noticed that many Christians in the West have an abundance of material possessions, yet they live in a backslidden state. They have silver and gold, but they don't rise up and walk in Jesus' name. In China few of us have any possessions to hold us down, so there's nothing preventing us from moving out for the Lord. We are praying that God

> *will use the Chinese church to help the Western church wake up and walk in the power of the Holy Spirit. It's almost impossible for the church in China to go to sleep in its present situation. There's always something to keep us on the run, and it's very difficult to sleep while you're running!" (pg 89)* [153]

Numbers are "approximate." – As you can see, the number of Chinese Christians outnumber Germany by approximately 5 to 1 today, and if trends stay consistent, they will be just under the United States' population by 2030. (Now is the time for the United States to "wake up;" as well as invest heavily on supporting the Asian, African, Latin American, and persecuted churches with at least 20% of our incomes, (as well as church offerings)). Whether we like it or not, they are the future of Christianity. *(See Chapter 3 and Phillip Jenkins' book "The Next Christendom" for more information)*

> *"With the destruction of the biblical faith in God and of all divine commands and ordinances, man destroys himself. There arises an unrestrained viatlism which involves the dissolution of all values and achieves its goal in final self-destruction, in the void.*"[52]

This "void" is the conclusion to complete abandonment of God, and moreover of Christ. Atheist philosopher Friedrich Nietzsche predicted that if people ever came to realize the full implications of their atheism, that this would usher in a total "void" of all meaning, values, and purpose. Nietzsche goes on to write that the end of Christianity means the ushering in of nihilism. But what we must ask ourselves, atheism/secularism/materialism (whatever you want to call it), have any validity? Should Christians be afraid to defend their faith? The answer to both is an overwhelming "No!" If I can understand these basic points and be able to convey them to either a novice or an expert, then anyone can.

Bonhoeffer goes on to describe an ever growing "western hostility towards the church..."

"pg 109 – The west is becoming hostile towards Christ. This is the peculiar situation of our time, and it is genuine decay. The task of the church is without parallel. The corpus christianum is broken asunder. The courpus Christi confronts a hostile world. The world has known Christ and has turned its back on Him, and it is to this world that the Church must now prove that Christ is the living Lord. The more central the message of the church, the greater now will be its effectiveness.[52]*"*

We have seen that this statement is ever increasingly true. Just this year I have seen a high school valedictorian's microphone muted when she tried to give thanks to Christ, (she was told she could say God, but not Christ), we also saw UCLA tell their graduates that they could say God, but not Christ. Not only is this becoming beyond ridiculous, every branch of knowledge we have access to, point towards a creator; and moreover leads us to Christ. We are seeing this ever increasing "educational persecution" throughout western schools and colleges. But as we will touch on briefly in the next chapter, this too is based on ideology and not evidence.

As I have mentioned, I would like to reiterate the fact that of the approximate 70 million martyrs since Christ, that approximately 45.5 million (65%), have came in the last century. This surprises many, but what we should all take notice of is it is my utmost opinion that as the saying goes "those who do not learn from the past are doomed to repeat it;" we should see that the 3 largest factors that have worked against

Christ centered Christianity, are the same 3 that are working the most against it at the present, and moreover, this will play the course of its future as we enter into the 22nd century. Please learn from our past, educate your church and church leaders about the persecuted church, and that these 3 factors are working right now at the present, and will undoubtedly affect our future into the 22nd century and beyond.

- Over 30,000,000 Christians have been martyred by atheistic powers in the last 100 years.
- Over 9,000,000 Christians have been martyred by Islam since its founding, (excluding Asia Minor estimates of another 40,000,000+) – an estimated 160,000 Christians have been martyred every year since 1990)
- Christians killing Christians – over 5.5 million [53]

These numbers are beyond alarming, and indeed tragic. No one can say persecution is not active today; nor is it not adequately addressed in the Bible and by Christ Himself. Let us always remember as Paul mentioned in Hebrew 13:3: "Remember those in bonds, as if you were bound with them." It is easy for us to remember those, but harder to do it "as if bound with them." I would encourage each and every one of us to actively seek ways to get our church involved and educated on the plight of our Brothers and Sisters in Christ who are dying every day; and to better prepare ourselves with the same courage and conviction they have to suffer and die before denying Christ. To educate yourself and others, (as well as use our affluence and wealth), to help and assist this body of Christ, while helping ourselves better grow in the process. The Christian church remained mostly silent against the Jewish genocide, and the German church as a majority chose to conform and compromise with political correctness and safety; will our approach be any different in the present or future? How much on persecution and martyrdom that occurs daily does your church talk about or speak out against? Food for thought - Please see the appendices for more information and what you can do.

> *"If we refuse to take up our cross and submit to suffering and rejection at the hands of men, we forfeit our fellowship with Christ and have ceased to follow him. But if we lose our lives in his service and carry our cross, we shall find our lives again in the fellowship of the cross with Christ. The opposite of discipleship is to be ashamed of Christ and his cross and all the offense which the cross brings in its train. Discipleship means allegiance to the suffering Christ; and it is therefore not at all surprising that Christians should be called upon to suffer."*
>
> *Dietrich Bonhoeffer, Cost of Discipleship, pg 91(martyred by Hitler in Nazi Germany)*

So what can we do about Islam and other ideologies within the United States?

I recently saw a YouTube video of Penn Fraser Jillette of Penn and Teller called "the gift of a bible," (see www.TheLollards.org or simply Google it or go to YouTube to watch). I did not know this person, but he is apparently a well known and ardent atheist and magic performer in Las Vegas. The YouTube clip of Penn is just him telling a story that summarizes a time following one of their Las Vegas shows when he met with some of the fans to sign autographs. Penn noticed a large man standing on the edge of the crowd. There is a segment in the show where they select an audience member who participates. This large man watching from the edge was an audience member during the previous night's show. He comes over to Penn and tells him how much he enjoyed the show. He's very complimentary Penn noted. The man looks him right in the eye, and tells Penn he has a gift for him. He presents him with a small Gideon bible. He tells Penn that he guesses he is proselytizing him but declares that he's not a crazy man; he's a successful business man who wants to give him this gift of a bible, a symbol of his faith.

Penn appears visibly taken by the man. "He looks me right in the eyes", says Penn, "and I know that he is not just leading me on or trying to butter me up when he tells me he likes the show." Penn says that he is an atheist and wonders out loud how much you have to hate someone not to tell them if you found something like God/Christ. He says that if we really believe in something that is so important why we would not tell them in the same way that we would push someone from the path of a truck if we could protect them from injury. "This is a kind, good man" said Penn. "He wants to give me a gift that means much to him. He's a good man."

That hit me hard. How much do we as Christians have to "hate" someone to not genuinely share our faith? I'm not referring to doing this in some creepy or fake way (as I will discuss in chapter 3), but in a genuine and sincere way. Likewise, how can we call ourselves Christ followers if we hate Muslims? Christ is our example and Christ loves Muslims so we should therefore love Muslims and everyone else if we are truly Christ followers. I didn't say love Islam, but to love Muslims. Moreover, if we genuinely love them, we will reach out to them as they are with the message of the Gospel of Christ. Not just in words, but in loving actions as Penn and Teller said that one individual showed to him, left a lasting impression. This means building relationships and removing barriers that are keeping them away from being a Christ follower; as I have told my good Muslim friend Nouran, I take no stock in "titles" so whether one refers to themselves as "Christian" is irrelevant to me. Following Christ and calling yourself whatever you would like if what is relevant to me, and hopefully to you as well. That is why it is a tragedy that we have not done more to reach out to our Islamic communities here in the United States; and why it is of the utmost importance to reach out to them now. I have made no qualm that as Ravi Zacharias has put it; fascist Islam is the biggest threat to the world and Christianity as a whole in the 22nd century and beyond. With that being said one might seem to think our plight is hopeless in reaching out to them, but this is false. As I have talked to many Muslims including my good friend Nouran, there are simple misconceptions held by most Muslims concerning Christ followers which can easily be answered and shown to be a simple misunderstanding. For example, Nouran once asked me how I could justify praying to a statue or idol; I quickly answered that I do not pray to idols and this is a sad path that some Roman Catholic churches hold to in appearance, but it is not biblical. After answering these questions and a few others on the Trinity and what not, the path was clear for me to steadily build a genuine relationship with her over 2009 and into 2010. To my surprise it is that easy. Just as at one time in history, no one thought that the war driven Vikings could ever be reached with the Gospel message, so too do many say that fascist Islam can never be humbled by the non-warlike Christ. Both assumptions are wrong. The Vikings were eventually reached with Christ, so too will the Muslims.

I must caution the reader - I am quite empathetic with Muslims or anyone else simply following Christ and not "westernization" of Christ. Just as Samuel Zwemer (better known as the "Apostle to Islam"), adhered to this, so I too must strongly encourage you to do as well.

> *"Zwemer's approach focused on language, literature, and scholarship. Zwemer devoted himself to the task, and he was later called upon to lecture and preach not only in his native English and Dutch, but also in the language of the Muslims to whom he was called. He founded the respected journal 'The Muslim Word' which is still published today, wrote and distributed numerous books and articles aimed at bridging gaps of understanding between Christians and Muslims, and labored tirelessly to mobilize a generation of Christians to engage Muslims peacefully. Living and traveling throughout the Arabian Peninsula and the entire Muslim world for decades, often under the worst of circumstances, Zwemer modeled the qualities of persistence and personal sacrifice (he buried 3 of his children in Arabia) that led eminent historian Kenneth Scott Latourette to state, 'No one through all the centuries of Christian mission to Muslims has deserved better than Dr. Zwemer the designation of Apostle to Islam.'"* [184]

Do we hate the Muslim, atheist, communist, fake Christian, etc, so much as Penn and Teller comments, that we will not reach out to them? Please do not get caught up on church/denomination/or even title, or even that they need to give up the mosque for a church; instead keep the focus on following Christ the savior, not Christ the prophet as the Koran proclaims. Leave the rest of the details to God. If the Holy Spirit inclines them to join your church, so be it; but do not let your own pride get in the way of Christ.

I am in no way concerned with my Muslim friend Nouran adopting the title "Christian;" my concern lies with her following and accepting Christ. God works in powerful ways, and has set the path before me to begin studying the Arabic language with Nouran and prepare to move to New York City where a large concentration of Muslims lives. We will see if my hypothesis of reaching out to the Muslim community is true or false. I earnestly pray that more will follow in Zwemer's footsteps now, so that the seeds planted will reap a great harvest of Muslims truly following Christ into the 22nd century and beyond.

*See Appendix D for more information on reaching out to Muslims and a free package to get you started.

A Celtic Cross on the Orkney Island in Northern Scotland, where the Vikings once dwelled and conquered. Just like the Vikings were eventually conquered by the love of Christ; so too can we conquer fascist Islam with dedicated/true love of the Muslim in the name of Christ.

What Could We Ever Learn from Persecution?

"Beyond Opinion" is an absolutely fantastic book by Ravi Zacharias and others, which has a chapter that I think answers the question of "what we can learn from persecution" quite well. Chapter 12 "The Role of Doubt and Persecution in Spiritual Transformation" is by Stuart McAllister, who was imprisoned on several occasions for distributing Christian literature and preaching the gospel in communist countries, recounts some of his reflections while in a communist prison:

> "I can well remember a point of surrender. After several days, I resigned myself to the possibility that my imprisonment could last for years. I might not get out for a long time, so I had to make the best of what was and to rest in God. It is a point where we accept the hardship, where we still believe in greater good, and where we surrender to what seems like inevitability. I think I came to relinquish my sense and need for control and simply accept that God would be there as promised, and therefore, to rest in him. I had crossed an important point that I subsequently discovered in the writings of Dietrich Bonhoeffer, Richard Wurmbrand, Alexander Solzhenitsyn, and Vadav Havel. First, I learned the role of prayer. I found that prayer is an active, ongoing, and vital conversation with God in the midst of struggle and doubt. Second, I learned the role of reflection. I thought about the great stories of the Bible and God's promises. In this, my memory of Scripture, songs, testimonies, and promises was crucial. What did they mean, and how did they apply here and now? Finally, I learned the role of struggle. As much as I disliked it, there was no

denying that struggle was all through the Bible, in the life of Jesus, and across church history." [169]

One story that resonated both with Stuart and my good friend Steven Khoury, (who is a Palestinian pastor in Bethlehem who also faces persecution on a daily basis); is the story found in the third chapter of the book of Daniel concerning Shadrach, Meshach, and Abednego. "The inauguration of the first Gentile Empire of Nebuchadnezzar's vision was marked by the enforced public worship of a golden image created by King Nebuchadnezzar. The king's agenda was to please his gods in fear that their anger would befall him and his people. This story tells us that when the trumpets and music were sounded every human being in the empire bowed down except for three young Jewish men who stuck out like sore thumbs. Imagine thousands of heads and knees bowing to these gods, probably including the king, and seeing these three men staring and watching the people around them. This was a strong testament to the truth they held so dear to their hearts. They could have decided to go along with everyone else and obey the commandment of the king, but they knew this was hypocrisy and a betrayal to Jehovah. I could even imagine other Jewish people or friends telling them it was not worth getting killed over or drawing attention to themselves, that God would understand their situation. This sounds a lot like what I have heard Christians say both in the Middle East and the West. The Jewish men wanted to stand firm in what they believed and were not afraid to let the whole nation know that they stood for their faith." [170]

"Shadrach, Meshach, and Abednego learned that this is a God-created and a God-governed world. Because it is created, they knew that living in conformity to God's will was the only way to truly function. Their view of reality was fuller that that of Nebuchadnezzar, and because of this they were confident that <u>doing</u> the right thing <u>was</u> the right thing. Because God governs the world, they knew that justice was ultimately in God's hands, not the king's. They were, therefore, willing to make a hard choice no matter where it would lead."[171] What about you and I? Would we be willing to face the "fiery furnace?" What about something much more close to home; such as losing your job, your car, some of your friends, maybe even tax-exempt status for your church? Are you or your church prepared? Pastor Wurmbrand said that a complete clamp-down of persecution happened in one day in Romania and we're seeing an ever increasingly hostile world to anything "God/Christ" oriented. As Pastor Wurmbrand stated in *The Triumphant Church*: "I would urge you to ask your synod or denomination to introduce courses on the underground church. Preparation for underground work begins by studying sufferology & martyrology. It is too difficult to prepare yourself for it when you are already in prison. You have to prepare yourself before hand for all eventualities. We have to prepare for suffering."[172]

So what do you learn? Stuart McAllister concludes: "I learned several things through facing persecution, struggle, and doubt. I needed to face my inadequacies and work at what Scripture calls the renewing of my mind. (Rom 12:2; Eph. 4:23) I needed to explore other ideas, worldviews, and systems and see how I could respond effectively to their claims. I needed to address my own spiritual needs and work on prayer, trust, self-awareness, and whatever disciplines were necessary to shore up my heart. I learned that because I did not have an answer or could not give an answer, it did not mean that an answer did not exist. I learned that the Christian life is reasonable, it is sufficient, but it is not safe, (if we mean a guarantee of problem-free living). Reality as I had understood it was 'other' than I thought. It was not just my journey from an unbelieving, atheistic worldview to a Christian one; it has been progressive exploration, correction, and transformation of what was often an idealized and romanticized Christian one. Reality, I have learned, is harder, more difficult, more demanding, and more complex than I was willing to admit or embrace. My illusions, false beliefs, and sincere but wrong thinking have been challenged and changed, often in the furnace of adversity."[173]

Alfredo Cerna (center) – Friend, teacher, pastor and fellow outreach partner who knows what it means to be persecuted. Pastor Cerna faced the atheistic communist regime in Nicaragua for part of the 1980's, and knows first hand as to the dangers of a fascist ideology.

What would Jesus say about persecution?

When I speak at churches, I begin many times by doing a simple quiz. One of the questions I ask is how many of the Apostles were martyred. The majority seems to not know this answer, but this is ultimately important to understand our history, and what Dietrich Bonhoeffer calls, "Costly Grace," (vs. cheap grace); all Apostles except for John who was exiled to Patmos where he wrote Revelation, were martyred for following Christ. Moreover, this is repeatedly told of by Christ:

"Remember the word that I said to you, 'A slave is not greater than his master.' If they persecuted Me, they will also persecute you; if they kept My word, they will keep yours also." John 15:20 (NASB)

The persecution of Christians is the religious persecution that Christians have endured as a consequence of professing their faith, both historically and in the current era. In the two thousand years of the Christian faith, about 70 million believers. Should we be surprised by this history of Christianity? It depends on how you look at it. Should we be surprised to see so much butchery in the name of Christ? Supposed Christians killing fellow Christians, Jews, Muslims, etc? Yes, we should be surprised and ashamed. This is nowhere taught by Christ. But should we be surprised at the hatred and persecution against Christians? Not at all.

Christ said plainly that darkness hates the light, and will seek to put out that light. The Bible also tells us plainly that our past, present, and future on earth will be marked by this persecution. It may seem surprising to most, but countries that suffer persecution seem to grow more than those secular countries that are spoiled and watered down. Did Pastor Wurmbrand's faith decrease for spending 14 years in prison, being beat and tortured for following Christ? No. As he said on page 63 of his book Tortured for Christ: "A flower, if you bruise it under your feet, rewards you by giving you perfume. Likewise Christians, tortured by the Communists, rewarded their torturers by love." Moreover, Christ and the Apostles words should serve as a wake up call to us, and for our action for those being killed right now in one of the many countries in the world which are hostile to Christ, Christianity, and Christians:

"Anyone who does not carry his cross and follow me cannot be my disciple… Any of you who does not give up everything he has cannot be my disciple." Luke 14:27,33

Blessed are those who are persecuted for righteousness' sake, for theirs is the kingdom of heaven.

"Blessed are you when others revile you and persecute you and utter all kinds of evil against you falsely on my account. Rejoice and be glad, for your reward is great in heaven, for so they persecuted the prophets who were before you.
—Matthew 5:10-12

"If the world hates you, know that it has hated me before it hated you. If you were of the world, the world would love you as its own; but because you are not of the world, but I chose you out of the world, therefore the world hates you. Remember the word that I said to you: 'A servant is not greater than his master.' If they persecuted me, they will also persecute you. If they kept my word, they will also keep yours.
—John 15:18-20

Who shall separate us from the love of Christ? Shall tribulation, or distress, or persecution, or famine, or nakedness, or danger, or sword? As it is written,

"For your sake we are being killed all the day long; we are regarded as sheep to be slaughtered."

No, in all these things we are more than conquerors through him who loved us. For I am sure that neither death nor life, nor angels nor rulers, nor things present nor things to come, nor powers, nor height nor depth, nor anything else in all creation, will be able to separate us from the love of God in Christ Jesus our Lord.
—Romans 8:35-39

We are afflicted in every way, but not crushed; perplexed, but not driven to despair; persecuted, but not forsaken; struck down, but not destroyed; always carrying in the body the death of Jesus, so that the life of Jesus may also be manifested in our bodies. For we who live are always being given over to death for Jesus' sake, so that the life of Jesus also may be manifested in our mortal flesh.
—2 Corinthians 4:8-11

For the sake of Christ, then, I am content with weaknesses, insults, hardships, persecutions, and calamities. For when I am weak, then I am strong.
—2 Corinthians 12:10

Indeed, all who desire to live a godly life in Christ Jesus will be persecuted,
—2 Timothy 3:12

Beloved, do not be surprised at the fiery trial when it comes upon you to test you, as though something strange were happening to you. But rejoice insofar as you share Christ's sufferings, that you may also rejoice and be glad when his glory is revealed. If you are insulted for the name of Christ, you are blessed, because the Spirit of glory and of God rests upon you.

Yet if anyone suffers as a Christian, let him not be ashamed, but let him glorify God in that name.
—*1 Peter 4:12-14, 16*

(Above quotes from English Standard Version copyright © 2001 by Crossway Bibles)

So I do pray that you will begin your own research on today's persecuted church, to keep them in your prayers, and to get involved in your own ministry today to alert the world around you to their plight. If a poor Romanian pastor who returning from 14 cruel years in prison, can begin a world wide ministry for the persecuted with an old type writer, then imagine what you could do.

[While I could go into details about specific countries and their plight, I will instead encourage you to research this information for yourself, by others who are much more qualified than I am, to speak of such matters. While I have had very little persecution, there are many others like Richard Wurmbrand, who can tell you first hand accounts, show the scars they have received, for following Christ. I highly recommend that you read Foxe's Book of Martyrs edition by Voice of the Martyrs (www.persecution.com), to get a better insight of this history and sign up for their free newsletter to keep you informed; as well as a simple search of the internet for Christian persecution. (Sadly you will find page after page that never makes it into the news headlines.)]

This brings us up to the present year of our Lord, 2009. Now that we have had a brief summary of where Christianity has been; let us see what the present state of Christianity is, and review (briefly), what every aspect of science, philosophy, and history can tell us concerning God and Christ, and how easily you too, can reach out to a skeptic or inquirer, (as well as insist at the very least, that our schools and universities are teaching "truth" over ideology.) Please note that these three simple chapters should at least begin to allow us to come full circle around a two-sided mirror, where we reflect on what being a "Christ-ian" really means, and for us all to ponder the question: "Am I actually a Christian?" And re-access if we are taking this role seriously in our personal as well as public lives, churches, and worship. It is my humble prayer, that this chapter has at least drawn your attention to the existence of Christian persecution throughout our history, and throughout our present world. Though it does not make the headlines, it is happening on a daily basis to our Brothers and Sisters in Christ. Christ promised we would see persecution if we followed Him, so may we learn from fellowship with the persecuted church and be prepared for our own persecution when the time comes.

> *"But martyrdom is not merely something for those in some far-off lands to think about. Everyone who claims the name 'Christian' should be preparing his or her heart for potential martyrdom. This is not an optional preparation for only those who live in Third World countries or those who live at certain times in world history. Preparing for martyrdom has always been part of what it means to be a true Christian. Christianity is the only religion that has as its highest example a man who was tortured and put to death publicly. As Christians, we are his followers. Yet the concept of martyrdom is essentially a foreign one to most of us in our Western Christian culture. But in many parts of the world today, such as China, Pakistan, or the Middle East, those who choose to follow Jesus all realize*

that they are saying yes to potential martyrdom. This was also the case for Christians for the first three hundred years of church history. Persecutions and martyrdom were common, especially among those who assumed positions of leadership." [161]

My wife's picture of the Statue of Liberty taken in December 2009 – While it appears that my wife and I are destined to be living in the New York City vicinity for the immediate future, we will have a constant reminder from the statue, (which is not far from where the Twin Towers once stood), which represents "liberty and freedom." May we each contemplate and reflect on what this means and not take our blessings for granted; the majority of Islamic countries are not allowed many of the freedoms we possess, including the freedom of religion, the freedom to hear the facts and decide for themselves which path to take, and moreover many face the death penalty if they leave Islam. (We are also seeing a growing limit in the freedom of education – the freedom to take the evidence of any/

all fields of study to their logical conclusions – if their conclusions point towards a "Creator" or to the fact that Jesus of Nazareth may have truly been who he claimed to be.)

May we remember the persecuted Church and the lessons it teaches –

Last page of Richard Wurmbrand's book Tortured for Christ, page 150:

"When I was beaten on the bottom of the feet, my tongue cried. Why did my tongue cry? It was not beaten. It cried because the tongue and feet are both part of the same body. And you free Christians are part of the same Body of Christ that is now beaten in prisons in restricted nations, that even now gives martyrs for Christ. Can you not feel our pain? The Early Church in all of its beauty, sacrifice, and dedication has come alive again in these countries.

While our Lord Jesus Christ agonized in prayer in the Garden of Gethsemane, Peter, James, and John were a mere stone's throw away from the greatest drama of history – but they were deep in sleep. How much of your own Christian concern and giving is directed toward the relief of the martyr church? Ask your pastors and church leaders what is being done in your name to help your brothers and sisters in restricted nations around the world. In these countries, the drama, bravery, and martyrdom of the Early Church are happening all over again – now – and the free Church sleeps.

Our brethren there, alone and without help, are waging the greatest, most courageous battle of the twentieth century, equal to the heroism, courage, and dedication of the Early Church. And the free Church sleeps on, oblivious of their struggle and agony, just as Peter, James, and John slept in the moment of their Savior's agony.

Will you also sleep while your brethren in Christ suffer and fight for the gospel? Will you hear our message: 'Remember us, help us? Don't abandon us!'

Now I have delivered the message from the faithful, martyred Church – from your brothers and sisters suffering in the bonds of atheistic communism (and fascist Islam), and under attack across the world from Indonesia to Africa. Don't abandon them." (Tortured for Christ, pg 150)

CHAPTER II
☙❧

*"There are numerous proofs for God's existence...
But none for His nonexistence.[54]"*
(Richard Wurmbrand, Proofs of God's Existence,
pg 122)

<u>Unprepared Christian vs. Secular University</u>

The date is August 1996, and I have just turned 18 and am ready for anything! I have been hired by the University of Arkansas' Microbiology laboratory as a lab technician. I meet and become friends with a fellow lab tech, Jenna, who is an atheist. I have never really had a talk with an atheist on anything so I am not really prepared, but let's see how I do! - As we're working on some media to grow salmonella cultures for a particular test, our conversation falls upon "meaning of life."
"*James.... Do you honestly believe there is really a God???????*" Jenna asks. I quickly come back with a great little answer of: "*Uh... Of course. You don't?*" - "*No!*" she exclaims. "*How about you tell me why there is evil, who made God, about evolution, about things being explained by our professors as happening completely on their own without any type of God. Explain this to me!*" "*Hmmmm.....*"

Christianity in the 22nd Century

I thought. I quickly came up with the normal Christian response: *"well.... You should just believe."* Then our conversation was done. Woo hoo! Way to follow 1 Peter 3:15! Give myself a brownie! We're told to "always be ready to defend your confidence in God when anyone asks you to explain it," but am I wrong when I say that at least 95% of Christians have just about the same answer when faced with these type questions? Am I also wrong that another 95% of these Christians say that being able to defend your faith is not important? I beg to differ. It is said approximately 2/3 of young Christians leave their church within their first two years of college, and our churches have done little about it.

I am horrible at math, but even I can figure out that these odds point to an ever increasingly shrinking church. (Example: Europe) -So why wouldn't this be one of the most important points for all of our churches to tackle? It's one of the biggest problems in the secular world, it is a biblical mandate, and we see Christ, the Apostles, and especially Paul following this example of Apologetics (from the Greek Apologia which means "to make a defense of"). So why wouldn't we have a Christian apologetics study in every church? Why is it that churches with an apologetics base are growing profoundly? *(See appendix B)*

Whether we like it or not, the church must remain progressive with needs. Take an example of the "computer." How many Christian organization and churches use computers? There are not too many church organizations that say *"No! We can not use computers! It isn't biblical!"* (Well, I'm sure there are a few...) - Of course they have not! They have remained "progressive" with the technology in that regard, so I am literally floored when a fellow Christian says that "knowing how to reach the skeptic is unimportant." (If I say, "should we follow 1 Peter 3:15, they will say 'of course!' (ironic, huh))

As I have stated, I am a very "average Joe," so if I can understand apologetics and reach out to attempt to answer a skeptics' questions, then anyone can. So premise two of this book is a genuine plea to our churches and fellow Christians, from a semi-young adult who cares greatly about our youth, to be able to "give a reason for your faith to anyone who asks you..."

As we mentioned in the last chapter, during the 18th century's Age of Enlightenment, atheism and agnosticism became widespread in Western Europe; 19th century Orientalism contributed to a certain popularity of Buddhism, and the 20th century brought increasing syncretism, New Age and various new religious movements divorcing spirituality from inherited traditions for many Europeans. The latest history brought increased secularization, and religious pluralism.[55]

Eurobarometer Poll 2005			
Country	Belief in a god	Belief in a spirit or life force	Belief in neither a spirit, god or life force
Spain	59%	21%	18%
Austria	54%	34%	8%
Lithuania	49%	36%	12%
Switzerland	48%	39%	9%
Germany	47%	25%	25%
Belgium	43%	29%	27%
Finland	41%	41%	16%
United Kingdom	38%	40%	20%

France	34%	27%	33%
Netherlands	34%	37%	27%
Norway	32%	47%	17%
Denmark	31%	49%	19%
Sweden	23%	53%	23%

The decrease in theism is illustrated in 1981 and 1999 according to the World Values Survey,[56] both for traditionally strongly theist countries (Spain: 86.8%:81.1%; Ireland 94.8%:93.7%) and for traditionally secular countries (Sweden: 51.9%:46.6%, France 61.8%:56.1%, Netherlands 65.3%:58.0%). Some countries nevertheless show a slight increase of theism over the period, Italy 84.1%:87.8%, Denmark 57.8%:62.1%. For a comprehensive study on Europe, see Mattei Dogan's "Religious Beliefs in Europe: Factors of Accelerated Decline" in *Research in the Social Scientific Study of Religion.*

Today, theism is losing prevalence in Europe in favor of Atheism and Humanism, and religion is losing prevalence in favor of secularism. European countries have experienced a decline in church membership and church attendance, as well as a decline in the number of people professing a belief in a god. A relevant example is that Sweden, which had 82.9% church membership as of 2000, had dropped to 72.9 % by 2008. However, in the 1990s, only 15% of the Swedish population said they believed in a personal God. It is generally thought that this disparity between church claims and numbers of people who actually believe in a god is likely to be the case in many other EU countries, especially in France and northern Europe, as recent trends and surveys are showing.[57]

These numbers are beyond troubling and confusing, but it gets even more confusing when you take secularism too far, (like much of Europe has). A large percentage of Europeans claim to be Christians, but also say they do not believe in God. How can this be? When I read this figure, I sat at my desk puzzled. I then began even more research not able to comprehend how someone can take a theistic religion (Christianity) that must presuppose deism (the belief in a God), and yet claim to be a Christian, but not a believer in God? This is honestly like saying, "I am a married bachelor," it is a contradiction in terms. So it amazes me that many of these persons who have made Christianity as either a type of pantheism or a type of social secular club, are the same people that look at true theistic followers of Christ as non-intellectual??? We will review this quickly from all fields of knowledge/research that we (humans) possess, and see which the most rational world view is; as well as show how you can easily handle this type questioner or skeptic in the future.

Many of our more militant atheists are coming from Europe, preaching their "fundamentalist secularism;" while at the same time, they are being enveloped by Islam, so the continent of Europe provides us with a good example to review, but hopefully not follow. Some of you might be asking what I mean by "militant atheist." Well, let's look at just a few quick examples:

> "If I could wave a magic wand and get rid of either rape or religion, I would not hesitate to get rid of religion."[58]

> "Daniel Dennett suggests in his book Darwin's Dangerous Idea, that religious believers who talk their children out of believing Darwinian evolution should be caged in zoos or quarantined because they pose a serious threat to the social order."[59]

Ouch! These are just two quick quotes from the new more outspoken atheists' generation, (most of their quotes are much worse). While most critiques for or against recognize their arguments, philosophy of science, and theological historicity as completely outdated and wrong; the fact that they say it with such unedited hate and conviction, has made many believer's and non-believer's alike shake with confusion and doubt.

The Bible and the Christians' role in Apologetics

"Education is thus a most powerful ally of humanism. What can a theistic Sunday school's meeting for an hour once a week and teaching only a fraction of the children do to stem the tide of the five-day program of humanistic teaching?"
(Humanism: A New Religion, 1930)

Once in awhile I hear from someone who is a fellow Christian saying something like: "we shouldn't have to give anyone any 'proof,' they should just believe!" While I agree 100%, that proof is not a substitute for faith in anyway, I can usually use any number of examples in the Bible to show that we are to be prepared to reach a skeptical world, but for the sake of time, I will just use one such example from the first actual "Christian Apologetic," Paul.

In Athens – Acts 17

[16]*While Paul was waiting for them in Athens, he was greatly distressed to see that the city was full of idols.* [17]*So he reasoned in the synagogue with the Jews and the God-fearing Greeks, as well as in the marketplace day by day with those who happened to be there.* [18]*A group of Epicurean and Stoic philosophers began to dispute with him. Some of them asked, "What is this babbler trying to say?" Others*

remarked, "He seems to be advocating foreign gods." They said this because Paul was preaching the good news about Jesus and the resurrection. [19]Then they took him and brought him to a meeting of the Areopagus, where they said to him, "May we know what this new teaching is that you are presenting? [20]You are bringing some strange ideas to our ears, and we want to know what they mean." [21](All the Athenians and the foreigners who lived there spent their time doing nothing but talking about and listening to the latest ideas.)

[22]Paul then stood up in the meeting of the Areopagus and said: "Men of Athens! I see that in every way you are very religious. [23]For as I walked around and looked carefully at your objects of worship, I even found an altar with this inscription: TO AN UNKNOWN GOD. Now what you worship as something unknown I am going to proclaim to you.

Paul did not simply say: "Everything you're doing is junk!" He first appealed to their reason and did not insult them, "Men of Athens! I see that in every way you are very religious." He then appealed directly to their culture and what they revered to get his foot in the door to share the Gospel: "I even found an altar with this inscription: TO AN UNKNOWN GOD. Now what you worship as something unknown I am going to proclaim to you."

In a world of growing skepticism and criticism, Christians need to humbly yet confidently be able to follow the guidance of the Holy Spirit, just like Paul did here, and Peter did when he tells us **Always** be prepared to give an answer to everyone who asks you to give the reason for the hope that you have. But do this with gentleness and respect.

I was asked to address our church on something dear to my heart, so I decided to talk about the importance of Christian Apologetics. Then I was quickly asked by the vast majority: "What are 'apologetics???'" So let's start from there – the word "apologetics" comes from the Greek word "apologia," pronounced, "ap-ol-og-ee'-ah." It means, "a verbal defense." And it is my humble opinion that every church should have a class or study designated to knowing just this; Christian apologetics. If our future depends on it, it is biblical and mandated by scripture, the "truth." Then what is the problem? (Please see Appendix B for more information)

What about you? Are you, your friends, kids, colleagues, church, etc, ready to handle attacks on your faith? Biblical doctrine says we should be, and many are surprised to find the vast majority of facts are of course on our side. Once again – this simple book is just to give you a snapshot spelled out by an average Joe, but it should be more than enough information to prepare you for 9 out of 10 encounters you run into; for the others please review my follow up information at the end of the book, for references books, websites, etc., to help you dive in more fully. So before we discuss the validity of Christ and Christianity, we need to be able to show someone that there is a "God." Let's first look at the area of science, and more importantly, philosophy of science and see where the evidence points.

"When interviewed for the 2005 National Study of Youth and Religion, thousands of teenagers who had been raised in religious homes said that over time, their faith slackened and they became "non-religious.[60]" Why? When asked to explain their loss of faith, their most common answer (32%) was intellectual skepticism. Specific answers included these: 'Some stuff was too far-fetched for me to believe in.' 'I think scientifically there is no proof.' 'There were too many questions that can't be answered.' Clearly, then, responsible discipleship today requires knowing the scientific evidence for an intelligent designer.[60]"

I could not agree more. With the evidence on the Christian's side, why would they not want to make stronger disciples before sending them off to secular schools?

So simply put, the two real hypotheses at debate from the onset, is between Naturalism/Darwinism (no Creator) and what is usually referred to as Intelligent Design (a Creator). Many Christians do not realize that we are living in an increasingly post-Christian United States, but we are also living in an increasingly non-God society all together. Therefore, we must educate ourselves to be able to reach those who do not believe in God, then be able to make a case for Christ (not church *per se*, but for Christ), and finally we must enable ourselves to truly have Christ in each of us. After all, if we are the body of Christ, we should always imitate Christ to the best of our ability; through our generosity, patience, love, and attitude.

So to begin with, let us please keep the two options of naturalism/Darwinism vs. Intelligent Design in mind while we briefly go through the key components of science and philosophy, and we will let you decide which is more plausible, and why it is important for us to endorse this type of simple approach, to show "both sides" to our students in both public education as well as private. Let's define each term and then simply look at what direction the evidence points. Intelligent design refers to a scientific research program as well as a community of scientists, philosophers and other scholars who seek evidence of design in nature. The theory of intelligent design holds that certain features of the universe and of living things are best explained by an intelligent cause, not an undirected process such as natural selection. Through the study and analysis of a system's components, a design theorist is able to determine whether various natural structures are the product of chance, natural law, intelligent design, or some combination thereof.

Such research is conducted by observing the types of information produced when intelligent agents act. Scientists then seek to find objects which have those same types of informational properties which we commonly know come from intelligence. Intelligent design has applied these scientific methods to detect design in irreducibly complex biological structures, the complex and specified information content in DNA, the life-sustaining physical architecture of the universe, and the geologically rapid origin of biological diversity in the fossil record during the Cambrian explosion. (Creationism is different in that while all Intelligent Design can get you to is a type of Deism, (that there is some type of being/first cause that began everything), while creationism is usually derived from the biblical accounts of theism.)[61]

Naturalism requires that its hypotheses be explained and tested only by reference to natural causes and events; that only the natural exists, (such as matter/materialism). Explanations of observable effects are considered to be practical and useful only when they hypothesize natural causes (i.e., specific mechanisms, not indeterminate miracles). Methodological naturalism is the principle underlying all of modern science. Some philosophers extend this idea, to varying extents, to all of philosophy too.

So unfortunately, what we are increasingly seeing is that regardless of any evidence, the secular agenda rules out any super natural cause regardless of any evidence, (which is not good/unbiased science).

It is my contention that both of the above theories should be taught at schools. I make no qualms about being a creationist myself, but I simply feel that all evidences should be taught at school and not hidden, regardless of whether they point to naturalism or theism; good or bad, just give us the facts and let us take them where the evidence goes. If we teach only one theory and exclude facts or questions about the theory, we are indoctrinating students, which I would think no one would agree with. That is why I feel that public schools' only option is to teach from an agnostic perspective, thus presenting all the facts. With that being said, let us look at the facts and let you decide for yourself which theory is the most logical. It is my hope that this chapter will simply give you a very basic overview of some of these facts and help to:

1) Strengthen your faith in God and Christ
2) Understand how to better reach out and answer the skeptic's questions/concerns
3) Get you informed to be able to reach out to your city officials, school board, senators, government officials, etc, on the importance of our students and faculty being able to take the evidence where it leads.

Let's start our search from the beginning, and see what astronomy, cosmology, and today's leading scientists say about the beginnings of the universe, space, and time itself.

The Cosmological Argument

> *"This religious faith of the scientists is violated by the discovery that the world had a beginning under conditions in which the known laws of physics are not valid, and as a product of forces or circumstances we cannot discover. When that happens, the scientist has lost control.*
> -Robert Jastrow (God and the Astronomers, 113-14)

Have you ever wondered where the universe came from? Why everything exists now as it is, instead of just nothingness? Most people do not realize, (or want to acknowledge), that the last 100 years of scientific discoveries have pointed more to a Designer that ever before. For many millennia, many thought of our vast universe as eternal and uncaused, but since the 20th century that has been universally abandoned, in favor of a created universe. (I'm not sure if you caught that or not, but what this means is all space and time itself began in the finite past, (it had a beginning point.)) This cosmological discovery is now most often referred to as the Big Bang.

The Big Bang is the cosmological model of the initial conditions and subsequent development of the Universe that is supported by the most comprehensive and accurate explanations from current scientific evidence and observation. As used by cosmologists, the term *Big Bang* generally refers to the idea that the Universe has expanded from a primordial hot and dense initial condition at some finite time in the past (currently estimated to have been approximately 13.7 billion years ago[62]), and continues to expand to this day. (This discovery alone verifies that the universe, space, matter, and time itself has an instantaneous beginning point.)

The framework for the model relies on Albert Einstein's general relativity and on simplifying assumptions (such as homogeneity and isotropy of space). The governing equations had been formulated by Alexander Friedmann. After Edwin Hubble discovered in 1929 that the distances to far away galaxies were generally proportional to their redshifts, as suggested by Lemaître in 1927, this observation was taken to indicate that all very distant galaxies and clusters have an apparent velocity directly away from our vantage point: the farther away, the higher the apparent velocity.[63] If the distance between galaxy clusters is increasing today, everything must have been closer together in the past.

Independently deriving Friedmann's equations in 1927, Georges Lemaître, a Belgian physicist and Roman Catholic priest, predicted that the recession of the nebulae was due to the expansion of the Universe.[64]

In 1931 Lemaître went further and suggested that the evident expansion in forward time required that the Universe contracted backwards in time, and would continue to do so until it could contract no further, bringing all the mass of the Universe into a single point, a "primeval atom", at a point in time before which time and space did not exist. As such, at this point, the fabric of time and space had not yet come into existence.[65]

What does this mean??? It means that science itself has confirmed the universe and time did have a first cause. (See appendix E for age of universe)

The cosmological argument is an argument for the existence of a First Cause (or instead, an Uncaused cause) to the universe, and by extension is often used as an argument for the existence of an "unconditioned" or "supreme" being, usually then identified as God. It is traditionally known as an argument from universal causation, an argument from first cause, the causal argument or the argument from existence.

This allows us to formulate the three following points:

1. Whatever begins to exist has a cause.
2. The universe began to exist.
3. Therefore the universe has a cause.

The implications of this finding are enormous, and unfortunately the ramifications are rarely spelled out to students. Even though it does not necessarily stipulate the God of the Bible, it does strongly support the Intelligent Designer of the cosmos, (and moreover it does match well with the creation story of Genesis).

I think agnostic Robert Jastrow who sat in the same chair as Edwin Hubble sums it up best in his book God and the Astronomers:

> *"It is not a matter of another year, another decade of work, another measurement, or another theory; at this moment it seems as though science will never be able to raise the curtain on the mystery of creation. For the scientist who has lived by his faith*

> *in the power of reason, the story ends like a bad dream. He has scaled the mountains of ignorance; he is about to conquer the highest peak; as he pulls himself over the final rock, he is greeted by a band of theologians who have been sitting there for centuries."*[67]

The discovery that the universe had a beginning was not met with pleasure. Many scientists rebelled against the notion because it implied a Beginner. One scientist admitted, "the notion of a beginning is repugnant to me." Yet the evidence was there. Jastrow puts his finger on the problem: Many scientists have a "religious" commitment to the assumption that everything has a natural, scientifically accessible and quantifiable explanation. Just when they were becoming confident in this assumption, seemingly explaining everything from the formation of stars to the formation of species, they ran into something which in principle cannot be explained scientifically: that first instant of creation, when the universe began as a singularity, a point inaccessible to investigation.

I feel I have said enough about this particular topic without going into too much complexity. I would highly recommend reading some of the books by William Lane Craig on this subject, that dive deeply into the cosmological argument. We will simply leave it at this, both believer and nonbeliever recognize that the universe did have a first cause, and a type of creation where not only space and matter, but time itself came into existence by this first cause.

The Teleological Argument

The next area we'll quickly look at is the fine tuning of the universe. If it all exploded from this finite singularity point, then it is highly unlikely that everything came together the way it needs for us to even be able to exist it would seem.

Did we simply get lucky as the naturalist would have us to believe? Why is the universe fine-tuned, and what exactly does this mean?

The fine-tuned Universe is the idea that the conditions that allow life in the Universe can only occur when certain universal physical constants lie within a very narrow range, so that if any of several fundamental constants were only slightly different the universe would be unlikely to be conducive to the establishment and development of matter, astronomical structures, elemental diversity, or life as it is presently understood.[68] Christian philosopher Alvin Plantinga argues that "random chance," applied to a single and sole universe, only raises the question as to why this universe could be so "lucky" as to have precise conditions that support life at least at some place (the Earth) and time (within millions of years of the present).[69]

This apparent fine-tuning of the universe is cited[70] by theologian William Lane Craig as an evidence for the existence of God or some form of intelligence capable of manipulating (or designing) the basic physics that govern the universe. Craig argues, however, "that the postulate of a divine Designer does not settle for us the religious question."

Naturalistic Darwinists choose to default to their favorite explanation that the entire fine tuning is simply due to "chance." But is this logical, or the best hypothesis? Just to give you a simple comparison of what the majority of people mean by chance, let me explain. We use chance everyday in someway. For example, if we hear on the weather forecast that there is a 90% chance of rain vs. a 10% chance of no rain, it is a fairly high probability that it will rain. Science likewise calculates such subjective statistics to determine whether something might or might not occur.

In day to day life, we can say that an event that has only one chance of occurrence in 10^8 (100 million) is considered an impossibility; to break it down, if you picked up a rock and dropped it 100,000,000 times and each time it fell to the ground, we have established that gravity is very highly probable. In scientific evaluations the number goes up to 10^{15} (quadrillion). When it comes to stating a scientific law, we are told that the number is 10^{50}. In other words if there is a mathematical probability of something occurring 10^{50} times, with only one chance of failure, the event is said to have been established by law, meaning it will always occur. Conversely, if there is only one chance of occurrence and 10^{50} chances of failure, the event is considered to be utterly impossible.[71] (Keep these probabilities in mind)

Just to put this in concrete terms, if I were to write the number 1 on a piece of paper every second 10^{20} times, it would take me 1.5 trillion years, (100 times the age of the entire universe); so I think it is safe to say these are huge numbers.

According to growing numbers of scientists, the laws and constants of nature are so "finely-tuned," and so many "coincidences" have occurred to allow for the possibility of life, the universe must have come into existence through intentional planning and intelligence. In fact, this "fine-tuning" is so pronounced, and the "coincidences" are so numerous, many scientists have come to espouse "The Anthropic Principle," which contends that the universe was brought into existence intentionally for the sake of producing mankind. Even those who do not accept The Anthropic Principle admit to the "fine-tuning" and conclude that the universe is "too contrived" to be a chance event.

The degree of fine-tuning is difficult to imagine. Dr. Hugh Ross gives an example of the least fine-tuned of the above four examples in his book, *The Creator and the Cosmos*, which is reproduced here:

One part in 10^{37} is such an incredibly sensitive balance that it is hard to visualize. The following analogy might help: Cover the entire North American continent in dimes all the way up to the moon, a height of about 239,000 miles (In comparison, the money to pay for the U.S. federal government debt would cover one square mile less than two feet deep with dimes.). Next, pile dimes from here to the moon on a billion other continents the same size as North America. Paint one dime red and mix it into the billions of piles of dimes. Blindfold a friend and ask him to pick out one dime. The odds that he will pick the red dime are one in 10^{37}. The nature of the universe reveals that a purely naturalistic cause for the universe is extremely unlikely and, therefore, illogical. One cannot say that a miraculous naturalistic event is a scientific explanation. Miracles are only possible when an immensely powerful Being intervenes to cause them. The Bible says that the fear of the Lord is the beginning of wisdom, and that He created the universe. When a model doesn't work, scientists must be willing to give up their model for a model that fits the facts better. In this case, the supernatural design model fits the data much better than the naturalistic random chance model. [72]

Teleology (Greek: *telos*: end, purpose) is the philosophical study of design and purpose. A teleological school of thought is one that holds all things to be designed for or directed toward a final result, that there is an inherent purpose or final cause for all that exists.

Therefore another hypothesis is derived from a teleological argument, or argument from design, which is an argument for the existence of God or a creator based on perceived evidence of order, purpose, design, or direction — or some combination of these — in nature. The word "teleological" is derived from the Greek word *telos*, meaning "end" or "purpose". Teleology is the supposition that there is purpose or directive principle in the works and processes of nature. This argument was strongly supported by Plato, Aristotle, Cicero, Averroes, Thomas Aquinas, and many others, but

William Paley summed it up in a more modern translation of his book Natural Theology in 1802 when he used his watchmaker analogy. The watchmaker analogy, or watchmaker argument, is a teleological argument for the existence of God. By way of an analogy, the argument states that design implies a designer. The analogy has played a prominent role in natural theology and the "argument from design," where it was used to support arguments for the existence of God and for the intelligent design of the universe.[73]

The watchmaker analogy consists of the comparison of some natural phenomenon to a watch. Typically, the analogy is presented as a prelude to the teleological argument and is generally presented as:

1. The complex inner workings of a watch necessitate an intelligent designer.
2. As with a watch, the complexity of X (a particular organ or organism, the structure of the solar system, life, the entire universe) necessitates a designer.

In this presentation, the watch analogy (step 1) does not function as a premise to an argument — rather it functions as a rhetorical device and a preamble. Its purpose is to establish the plausibility of the general premise: *you can tell, simply by looking at something, whether or not it was the product of intelligent design.*

In most formulations of the argument, the characteristic that indicates intelligent design is left implicit. In some formulations, the characteristic is *orderliness* or *complexity* (which is a form of order). In other cases it is *clearly being designed for a purpose,* where *clearly* is usually left undefined.[73] Now let's fast forward to the 20th century and what we see is the teleological argument taken to an unimaginable level.

"For since the creation of the world His invisible attributes, His eternal power and divine nature, have been clearly seen, being understood through what has been made, so that they are without excuse."
Romans 1:20

A modern variation of the teleological argument is built upon the anthropic principle. The anthropic principle is derived from the apparent delicate balance of conditions necessary for human life. In this line of reasoning, speculation about the vast, perhaps infinite, range of possible conditions in which life *could not* exist is compared to the speculated *improbability* of achieving conditions in which life *does* exist, and then interpreted as indicating a fine-tuned universe specifically designed so human life is possible. This view is shared by John D. Barrow and Frank J. Tipler in *The Anthropic Cosmological Principle* (1986).

The extent of the universe's fine-tuning makes the anthropic principle one of the most powerful arguments for the existence of some type of creator. There are more than 100 constants that strongly point to an Intelligent Designer, so we are just scratching the surface, but one example would be if gravity was 0.00000000000000000000000000000000 0001 percent altered, our sun would not exist, and, therefore, neither would we.[164] (And this is just ONE constant of many)

As Discover magazine's November 2002 issue put it: "The universe is unlikely. Very unlikely. Deeply, shockingly unlikely." (Brad Lemley, "Why Is There Life?")

So when there is virtually zero chance of a universe, (much less a life permitting universe), occurring by chance, how then can anyone still hang on to the chance hypothesis?

In his interview with physicist and mathematician Robin Collins (PhD), Lee Strobel sums it up:

"In light of the infinitesimal odds of getting all the right dial settings for the constants of physics, the forces of nature, and other physical laws and principles necessary for life, it seems fruitless to try to explain away all of this fine-tuning as merely the product of random happenstance.
'As long as we're talking about probabilities, then theoretically you can't rule out the possibility – however remote – that this could occur by chance,' Collins said. 'However, if I bet you a thousand dollars that I could flip a coin and get heads fifty times in a row, and then I proceeded to do it, you wouldn't accept that. You'd know that the odds against that are so improbable, that it's extraordinarily unlikely to happen. The fact that I was able to do it against such monumental odds would be strong evidence to you that the game had been rigged. And the same is true for the fine-tuning of the universe – before you'd conclude that random chance was responsible, you'd conclude that there is strong evidence that the universe was rigged. That is, designed. In a similar way, it's supremely improbable that the fine tuning of the universe could have occurred at random, but it's not at all improbable if it were the work of an intelligent designer. So it's quite reasonable to choose the design theory over the chance theory. We reason that way all the time. Were the defendant's fingerprints on the gun because of a chance formation of chemicals or because he touched the weapon? Jurors don't hesitate to confidently conclude that he touched the gun if the odds against chance are so astronomical.'"[74]

Are we beginning to see why Jesus said, [31]"He said to him, 'If they do not listen to Moses and the Prophets, they will not be convinced even if someone rises from the dead.'"

Dr. Subodh Pandit

I have had the honor of working with Dr. Subodh Pandit of India, in his Search for God seminars (www.SearchSeminars.org), since 2008. In our first meeting, we had several proclaiming atheist professors come to debate. One finally said, "Why doesn't God just spell in the stars the word 'BELIEVE,' then we would believe." But that is false. We would simply extrapolate a theory on how those stars "looked" designed, but somehow they naturalistically came together to "look" like the letters to spell "Believe;" and thus they would dismiss it as yet another naturalistic occurrence. In the same seminar, an atheist philosopher stood up when I commented on the chance of their "chance hypothesis" being so weak. His comment (from a PhD professor almost floored me); he said "I admit that the chances of something like Mt. Rushmore being formed strictly by chance over thousands of years of wind, rain, and erosion to look like the presidents is extremely improbable, but it is still a chance and God is no chance!" I was aghast; this was what his reason and logic concluded, and the reasoning behind it? (This type scenario should encourage all of you reading this to not be afraid of challenging these types of ideas in a public forum; if this is the best five atheistic professors can muster, than we need not fear a peaceful exchange of our ideas for theirs).

There's no plausible explanation for the anthropic principle other than for a Cosmic Designer; it would seem. Atheists must take extreme measures to deny the obvious. When they dream up theories that are not supported by any evidence – and in fact are actually impossible by their own standards – they have left the realm of reason and rationality and entered into the realm of blind faith.

Believing without observation is exactly what atheists accuse religious people of doing; but ironically, it's the atheists who are pushing a religion of blind faith. Believers have a good reason (such as the big bang and anthropic principle), for believing what they believe. Atheists really do not. This blind faith of the atheist reveals that the rejection of a Designer is not a head problem – it's not as if we lack evidence or intellectual justification for a Designer. On the contrary, the evidence is impressive. What we have here is a will problem – some people, despite the evidence, simply do not want to admit there's a Designer. Just as science has been hindered throughout its history by atheists, not wishing to acknowledge such events as the big bang because of its theistic ties, one such scientist spoke of the Anthropic Principle in the NY TIMES, OCT 2003 edition, and blatantly said his real objection was "totally emotional" because "it smells of religion and intelligent Design." - So much for scientific objectivity...

We now have an adequate amount of information to formulate the following:

1) The fine tuning of the universe is due to either law, chance, or design.
2) It is not due to law or chance.
3) Therefore, it is due to design.

Lee Strobel similarly asks Dr. Collins: "'What has your study of fine-tuning of the universe done to your faith?' I asked. Collins put down his tea. 'Oh, it has strengthened it, absolutely,' he replied. 'Like everybody, I've gone through some hard times in life, and all the scientific evidence for God has been an important anchor for me.' That sounded like science displacing faith. 'Isn't that what faith is supposed to do?' I asked. 'I am talking about faith,' he insisted. 'God doesn't usually appear supernaturally somewhere and say, 'Here I am.' He uses preachers to bring people his message of redemption through Christ. And sometimes he uses natural means. Romans 1:20 tells us that God's eternal power and divine nature can be seen and understood through things that are made, and that this is the reason humanity is without excuse. I see physics as uncovering the evidence of God's fingerprint at a deeper and more subtle level than the ancients could have dreamed of. He has used physics to enable me to see the evidence of his presence and creative ability. The heavens really do declare the glory of God, even more so for someone trained with physics and with eyes to see. That has been a tremendous encouragement to me.

Of course,' he continued, 'the fine-tuning by itself can't tell us whether God is personal or not. We have to find out in other ways. But it does help us conclude that he exists, that he created the world, and that therefore the universe has a purpose. He made it very carefully and quite precisely as a habitat for intelligent life. It's (the anthropic principle), not conclusive in the sense that mathematics tells us two plus two equals four,' he said. 'Instead, it's a cumulative argument. The extraordinary fine-tuning of the laws and constants of nature, their beauty, their discoverability, their intelligibility – all of this combines to make the God hypothesis the most reasonable choice we have. All other theories fall short.'"[75]

At this point, it is important to point out that if by some chance, all physics, math, and beyond all imaginable probabilities this all did happen by chance, all we have is an empty planet. A habitable fish bowl so to speak. But we do not have any life. This is even another "beyond possible" chance occurrence. We will briefly dissect what is needed now for life to somehow form by chance. (Once again, I will only be touching on this subject, so please reference the back of the book, (or on your own), for book after book on the subject).

The Origin of Life – Chance vs. God

So now... How did the first life come about? I will quote you from a major university's biology text book that I picked up at random what the naturalist states as fact to students:

"Origin of Life: Today we do not believe that life arises spontaneously from nonlife, and we say that 'life comes only from life.' But if this is so, how did the first form of life come about? Since it was the very first living thing, it had to come from nonliving chemicals."[76]

That's it? I may not be an expert in philosophy, but I am pretty sure those premises do not line up.

1) Today life does not come from non-life.
2) The first life was early.
3) Therefore the first life came from non-living chemicals.

How they were able to get premise 3 in the majority of university biology text books is mind boggling. This is a classic example of horrible logic and reasoning within a modern collegiate text book. Okay then - so we have heard the naturalist's explanation. (Chance again)

Let's quickly review the law they were referring to in the text book when they commented on "life comes only from life."

Law of Biogenesis

Redi's and Pasteur's theory that modern organisms do not spontaneously arise in nature from non-life is referred to as the *law of biogenesis*. Thomas Huxley made the declaration that biogenesis was indeed a law of nature in his 1870 address to the British Association for the Advancement of Science, where he summarized the long scientific debate of Abiogenesis vs. Biogenesis.[77]

So since this was the "first life," we can arbitrarily throw out scientific criteria?

Long story-short, evolution states that somehow in what they call the abiogenesis, or "chemical evolution," life on Earth could have arisen from inanimate matter. Two naturalistic theories are that there are two possible sources of organic molecules on the early Earth:

1. Terrestrial origins – organic synthesis driven by impact shocks or by other energy sources (such as ultraviolet light or electrical discharges)
2. Extraterrestrial origins – delivery by objects or gravitational attraction of organic molecules or primitive life-forms from space (aka – Space aliens)

Either by "chance" interaction of chemicals, or that "chance" delivery from space could have delivered the first beginnings of life. (I still am somewhat dumbfounded that these 100% speculative theories are even considered credible, while intelligent design is not.)

"Dr. Michael Denton is considered one of the first to statistically challenge Darwin's idea of spontaneous generation of life. He looked at the simplest cell that could possibly live and function. What were the chances that one hundred proteins, (the smallest number required), could come together by sheer coincidence? His calculation showed a probability of one in 10^{2000}. Remember that 10^{50} is an impossibility. Ralph Muncaster, in his book 'A Skeptic's Search for God,' stated that the chances of getting 10,000 amino acids with left-sided links (which is absolutely necessary), and 100,000 nucleotides with right-sided links together in one cell, was one in 10^{33113}. Harold Morowitz calculated the odds of a whole cell randomly assembling under the most ideal circumstances to be on in $10^{100,000,000,000}$! And we have not computed the other factors in, like the exact function of each molecule, division of that cell, increasing biological complexity, life from some unknown source injected into the organism, etc. If the list itself is endless, what would combined chances be for all components to come into existence and then come together spontaneously, by sheer coincidence? Statistically, an absolute impossibility."[78] So without going into tons of details on the first life piece, I will simply encourage you to research these options yourself as we dive into a few more details post the origin science of life.

William Dembski allows those in favor of "chance" an even larger area for "chance" to be a chance; but he still argues convincingly that there is a statistical limit to this improbability. Anything that is less than $1/1 \times 10^{150}$ is essentially statistically impossible within our existing universe. He arrives at this number by multiplying the total number of elemental particles estimated to exist in the entire universe (1×10^{80}) times the number of transitions that each elemental particle can make in a second (1×10^{45}) times the supposed age of the universe, assuming the universe is a billion times older than 20 billion years old (1×10^{25} seconds) = 1×10^{150}.

While this is a huge number, it is not large at all when contemplating very complex systems. Hubert Yockey, a highly regarded information theorist, has calculated the amount of information content in the minimum genome for life to arise and the probability of that occurring by chance as something less probable than $10^{186,000}$. (Hubert Yockey, Calculating Evolution, Vol. 3 No. 1, p 28 (Cosmic Pursuit, 2003)) So let us be honest for a moment – who has the most faith – the supporter of the God hypothesis, or the Chance hypothesis?[79]

The only real game in town for the naturalist is that of Darwinism, so it would be a good subject to spend a few pages on, so that you can see the argument and know how to reach out to a skeptic. What is Darwinian evolution in a nutshell? Is Darwinism and Intelligent Design science? Let's quickly look at each of these terms and leave the ball in your court to decide.

Evolution – Fact or Fiction?

Sci·ence (sī'əns)
n.
1.
a. The observation, identification, description, experimental investigation, and theoretical explanation of phenomena.
b. **Such activities restricted to a class of natural phenomena.**
c. Such activities applied to an object of inquiry or study.
2. Methodological activity, discipline, or study: I've got packing a suitcase down to a science.
3. An activity that appears to require study and method: the science of purchasing.
4. Knowledge, especially that gained through experience.
5. Science Christian Science.

Isn't it interesting how scientists say that they are "open" minded, but by the very definition of "science" they rule out any supernatural or metaphysical options. (Such activities restricted to a class of natural phenomena.) - Science at one point simply meant "knowledge," then it grew to include the scientific method, and as you can see, now it encompasses several different definitions – in which "B" rules out anything other than "natural phenomena." - This seems to be self-refuting of option # 1 (theoretical explanation of phenomena.) - Why is it that some of the most "free" nations in the developed west, have the least freedom of inquiry when it comes to educational freedoms? - As many of you have heard, there is an ever-increasing movement urging us to be able to simply follow the evidence where it leads, regardless of the implications, but governmental agencies are restricting this very option, which is at the heart of science, freedom, and education. I would think anyone; theist, agnostic, or atheist would welcome a true openness to all the facts and following the evidence where it leads, but this is becoming less and less so.

Charles Darwin himself said that all evidences should be given to the populace. (In my opinion, I truly think if Darwin were alive today, he would be a supporter of intelligent design theory.)

When my great – grandparents were children, they were taught in public schools from a more theist approach in public schools, my grandparents more of a deist approach, my parents more of an agnostic approach with a leaning towards theism, my generation saw a more straight agnostic approach, and now today's children are getting a more atheistic approach. This truly is not a warranted or scientific approach, but an ideological one, that unfortunately our elected officials and Christians as a whole, have let happen without much of a fight. So as the documentary film EXPELLED: No Intelligence Allowed showed, our society has an ever-growing intolerance to anyone that challenges the naturalistic/Darwinist approach to explain origins. As I had mentioned earlier, we are living in a post Christian America, but this is no reason for an inquiring mind, much less a Christian mind, to roll-over and play dead. I am honestly dumbfounded by much of the way this has fallen apart, and the inactivity that Christians are playing in it all.

We will first look at what evolution is and is not. One could break evolution into the following categories:

1) Cosmic Evolution
2) Chemical Evolution
3) Organic Evolution
4) Macro Evolution
5) Micro Evolution

Of these terms, only micro evolution, (change over time), has been verified. This term was never really controversial in the first place and existed well before Darwin. An example would be Eskimos having a higher fat content because they live in colder climates than someone who lives on the equator. There are some changes notable as the generations have gone on, but that is where it stops. We have big dogs and little dogs, but they are always dogs. This is an example of micro evolution, which no one to my knowledge contests.

The other theories are 100% speculative and must be followed by extreme faith. Macro evolution contends that at some point, everything can be traced back to a single cell that was created by complete chance as mentioned above. (Modern evolutionary trees have slowly "evolved" to include four super groups, (Plants, parasites, fungi/all animals, and algae related)). This is related to the term we hear called "common descent" which is generally accepted that all living organisms on Earth are descended from a common ancestor or ancestral gene pool. In "layman's terms," this basically means that somehow beyond all probability, life came about from what is often called "primordial soup" from the early Earth, and then it somehow came to life and basically went through a series of drastic evolutionary changes/mutations followed by natural selection to account for all the phylum, genus, species, etc., we have today.

There are an incalculable number of problems with this theory. First of all it can not be tested; therefore it sets itself up as un-falsifiable. We can not test origin science, so cosmic, chemical, and organic evolutions are 100% conjectured. We do not see any of these changes happening at present, (because it would take at least millions of years), we can not duplicate any of these experiments, so we are left with conjecturing and theorizing, but there is no data available for us to have any real confidence in these speculations. We also do not have any reason to believe that macro evolution is plausible. We have never seen this happen, (because once again, it would take millions of years), there should be a plethora of transitional intermediary fossils found, but there is not; moreover, the metamorphic changes required are grossly neglected by the vast majority of supporters to this idea. For example, the cow to whale transition is supposed to be one of the stronger evidences in favor of evolution, but mathematician David Berlinski has calculated at least 50,000 metamorphic changes would be required to go from the cow to the whale. We do not see any evidence for this today, we do not see any evidence for this in the fossil record, nor would the fossils even be able to give us a great picture because they are un-testable themselves. (In fact, we are not able to prove by looking at a seal fossil and a walrus fossil that one "changed" into the other). Once again, it is a speculation, which everyone is entitled to make; but this is a relatively weak theory when all that has been proven since well before Darwin, is that we see variations within genus. We have all types of different people, but we do not have half ape/half man creatures, nor do we have any reason to think there ever were such creatures. Think about it, we should not only have at least some species/genus of ape men still alive, but we should have a ton of fossils showing these changes.

Instead we see a history of more than 100 years of frauds and fakes trying to find just one single ape-man for Darwinists to get behind. I encourage you to research these on your own, but just to name a few of the famous fakes of history that still exist in some text books: Piltdown man, Nebraska man, Peking man, Neanderthal, Lucy, and the list goes on. While I was attending the University of Arkansas, I took a class in 2001 called Biological Anthropology, which focused primarily on human evolution. Neither could the professor answer any of these types of questions, but some of the criteria they mapped out for the class was changed a few years later because new fossil discoveries proved their evolutionary tree wrong, but it also caused scientists yet again to acknowledge the complexity of attempting to force evidence together that does not exist. It is hard for me to believe, when I see top scientists pick up four or five fossils and say that these prove something such as the cow to whale evolution, or that they find a single tooth, (such as Nebraska Man), and say this tooth proves that humans evolved from apes. (The tooth ended up being that of an extinct pig) The list continues to grow, with usually at least one headline per year exclaiming: "Missing link found for sure this time!"

Although Darwin was not the first to come up with his "Tree of Life," his was the one to receive the most notoriety for it focused on the second part of the definition of "science" – by natural means (excluding the supernatural). Even a source such as Wikipedia, shows that while Darwin's tree was not the most sound, it was considered an accurate tree because it excluded any type of supernatural element:

"The earliest tree of life was published by the French botanist Augustin Augier in 1801. It shows the relationships between members of the plant kingdom. It was not an evolutionary tree because a Creator was involved. Jean-Baptiste Lamarck (1744-1829) produced the first branching tree of animals in his Philosophy zoologique (1809). It was an upside-down tree starting with worms and ending with mammals. No Creator was involved, so it is an evolutionary tree. However, Lamarck did not believe in common descent of all life. Instead, he believed that life consists of separate parallel lines advancing from simple to complex[81]. The American geologist Edward Hitchcock (1763–1864) published in 1840 the first Tree of Life based on paleontology in his Elementary Geology[82]. On the vertical axis are paleontological periods. Hitchcock made a separate tree for plants (left) and animals (right). The plant- and the animal tree are not connected at the bottom of the chart. Furthermore, each tree starts with multiple origins. Hitchcock's tree was more realistic than Darwin's 1859 theoretical tree because Hitchcock used real names in his trees. It is also true that Hitchcock's trees were branching trees. However, they were not real evolutionary trees, because Hitchcock believed that a deity was the agent of change. That was an important difference with Darwin. In 1858, a year before Darwin, the paleontologist Heinrich Georg Bronn (1800-1862) published a hypothetical tree labeled with letters. Although not a creationist, Bronn did not propose a mechanism of change."[83]

As William Dembski and Sean McDowell brilliantly show in their book "Understanding Intelligent Design pgs 20-26[84]: "Darwinism is one of the few subjects that popular culture holds in awe. It is often treated as the pinnacle of science (an honor that we'll soon see is underserved...)

A GREAT example, in a television program called "Friends," had an episode that displayed Darwinism in society; Phoebe and Ross discuss the merits of Darwinian evolution. Shocked to find that Phoebe rejects it, Ross says: "Uh, excuse me. Evolution is not for you to buy Phoebe. Evolution is scientific fact, like, like, like the air we breathe, like gravity."

Are supporters of Darwinism giving us a fair picture here? Is it just plain silly to deny Darwinism? What if there is a better explanation for the origin and structure of the universe than Darwinian naturalism? What if the world has been designed for a particular purpose? If so, the attempt to understand all reality within a Darwinian perspective would be a colossal mistake.

Darwin's theory of evolution, often referred to as "Darwinism," inspires the best-known form of naturalism. In Darwin's Dangerous Idea, philosopher Daniel Dennett says that every aspect of human life – including education, relationships, and politics – must be life – including education, relationships, and politics – must be understood in light of Darwinian evolution. Dennett wants ALL reality to be understood within a Darwinian framework. Why? Because all Darwinism tells us our creation story, and such a story always controls how we interpret reality. Given how widely Darwinism is accepted, we should not be surprised that virtually every field of study is now being "Darwinized."

For half a century, Anthony Flew was the world's most famous intellectual atheist. Then in 2004, he made a shocking announcement: God must exist. In a headline-making reversal, Flew now holds that the universe must be the work of an intelligent designer.[84] In an interview for Philosophia Christi, he added, "It now seems to me that the findings of more than fifty years of DNA research have provided materials for a new and enormously powerful argument to design."[85]

More and more scientists agree with Antony Flew: The world appears designed because it is designed. They argue that the design in the world is just as real as the design in a computer chip, a car, or a sports stadium. These scientists have also observed another surprising thing: The hard empirical evidence for Darwinism is in fact very, very limited. Darwin's mechanism of natural selection acting on random variations can account for small-scale changes in living forms: variations in species themselves. But neither

Darwin's mechanism nor any other purely natural mechanism explains how insects and birds came to exist in the first place. The theory is supposed to explain such large-scale adaptations, but it doesn't and there has not ever been ONE SINGLE point of evidence other than pure speculation and guessing, so I at least, am utterly dumbfounded why after 150 years from the release of Darwin's book, that this theory is still taken seriously to the degree which it claims.

Having taken Evolution/Darwin at a major secular university, I can testify that while the evidence is not there, Darwinists would like to believe that people are skeptical of Darwin's theory only because they've been religiously indoctrinated and haven't been properly educated. But this is a curious claim since the past 40 years public schools have exclusively taught Darwinism. Rather than a problem of education, the problem is one of evidence: Darwinism simply doesn't have the goods. Despite confident assertions, the actual evidence in favor of Darwin's grand theory of evolution is surprisingly thin. <u>What Martin Luther helped accomplish in Augsburg to push open the door of religious freedom, now needs to be done in our generation, opening the door of educational freedom, not indoctrination; and to simply have the freedom to take the evidence where it leads.</u> [84] (see appendix B)

While we could dive into book after book on the theory of evolution, enough has been said to establish a solid base that evolution requires at least as much faith as does the hypothesis of a supernatural creator. However, I am by no means using a type of "God of the Gaps" theory that many will claim. I simply continue to reveal the hard core facts to both hypothesize, and challenge you to do the same, because none of the basic premises I am saying here are disagreed upon even by those who are Darwinists themselves. I am only using the most basic of facts without going into the minute details. This I am neither qualified for, nor seek to accomplish. I merely seek to review the topics discussed on a macro, easily accessible approach to give both layman and expert a presentation to sit back and contemplate. What we can establish, is that evolution does now qualify as an ideology and philosophy, more than it does as a credible interpretation of the facts. So while we can all agree with change over time, also known as micro evolution, any other type "evolution" past this point is speculative at best. When a group accepts all definitions of evolution discussed earlier as fact, and begin incorporating this in all of their various outlooks, this becomes an ideology that many now call "Darwinism."

What about "theistic evolution?"

Many churches are beginning to accept what is referred to as theistic evolution. This basically states that God started life, and then all of the evolutionary changes began and these changes eventually resulted in you and me being formed as we are today. We have already looked at the extreme improbability of any of the realms of science to account for evolution beyond simply changes over time, (micro evolution), so why would a Christian want to mix in a highly improbable theory with their godly/theistic beliefs? Moreover, most people do not think about it, but evolution is based off of naturalistic causes which rule out God or the supernatural in the first place. So how can we possibly attempt to use a strictly unguided process to describe how God guided the process? The two are completely contradictory in terms, and are therefore incompatible. We have already discussed why evolution is not a very probable explanation for how life developed from a non-theistic view; moreover, it is absolutely an illogical concept for one to claim a belief in bridging Darwinism with theism to get theistic evolution. Most supporters of this view that I have encountered, quickly abandon the view when we begin to discuss it, simply because they did not have a full grasp of what it implied, and how it contradicted not only probability, science as we know it, but also Christian theology. I read an article by a pastor that I thought summed it up well:

A Bad Foundation by Doug Bachelor (on The Bible and Evolution):
"Teaching our children the lesson that there is no absolute right and wrong is very dangerous. It has caused a disaster in our public schools, our court system, and for the very fabric of our society. A false understanding of human origins ultimately degrades society. Consider the nations that have made atheism the core of their culture — the former Soviet Union, Cuba, China, and Vietnam. I've been to Russia and China and have seen the devastating effects of atheism: Suicide, alcoholism, and spousal abuse are epidemic. Atheism offers no hope or purpose for life. But exhibit "A" would be the drastic differences between North and South Korea. If you stand on the 38th parallel, you'll see a very bleak and backward existence of the imprisoned people of the North. Look south toward Seoul and you will see a bright and free civilized existence. The core difference? South Korea is a Christian stronghold; North Korea teaches evolution and atheism. Satan told Adam and Eve that if they rejected God's Word, they would be freed and experience unlimited human advancement. Instead, they were enslaved by sin. Today, Cuba, North Korea, and China aggressively persecute Christianity, all the while suppressing freedom, advancement, and hope — enslaving their people in unspeakable evil. Evolutionists can also rationalize all kinds of immoral behaviors as merely part of the evolution of man; nothing is inherently bad. As a teenager, I learned that my science teacher had an affair with a woman in the loft of his home while his pregnant wife was downstairs.

Though it deeply hurt his wife, he appeared indifferent to her feelings. He excused himself by saying, 'Not all of the primates we've evolved from are monogamous, so adultery is perfectly natural. We can't help it.' Evolution clearly undermines Christian living."[146]

I think Doug sums it up well in this article. Not only is evolution not compatible with our knowledge of science, but it also takes the glory from God, and gives it to "chance." Chance is the atheists' God. Also we have to begin editing the Bible, and discrediting large portions of it. For example, were Adam and Eve both primates? Were they proteins in the primordial soup of chance, which ultimately gave us life? Or are they fully human, just like the Bible states, and the evidence that we have at our disposal, shows us? I am dumbfounded that so many think Darwinian evolution is compatible to account for the complexity of life, along with life itself; when it is 100% speculative and conjecturing of the evidence. Moreover, I am absolutely floored that theistic believers can somehow combine this theory with their theism, to come up with the term "theistic evolution." It is obviously an attempt to compromise atheism with theism, logic with illogic, reason with non-reason; but I think we can conclude that this type of thinking is completely unwarranted.

Was Adam a highly intelligent primate? Of all the questions that can be raised about the theory of evolution, none is more vital than whether humans evolved from non-human animals. In naturalistic evolutionism, human beings were not created with a dignity transcending all other animals, but instead are simply a particularly intelligent primate. The biblical teaching is that the human race has fallen from an original innocence, and that our tendencies to violence, greed, lust, deceit, and selfishness are in some sense unnatural for us. This teaching is at direct odds with the notion that the human race evolved from similar primate species, and that our unethical tendencies are actually part of our evolutionary history, (aka - "survival of the fittest").

In addition to the problems attending the general theory of evolution, the evolutionary explanation for the origin of the human species has been plagued by the question of the "missing link." In the first half of his book *The Bone Peddlers: Selling Evolution*, William R. Fix reviewed the history of frauds, hoaxes, and misidentifications that has characterized the search for the missing link between *Homo sapiens* and the lower primates from which we supposedly evolved. Two of the most notorious of these bogus links were Piltdown Man, a fraud constructed with sawed-off bones, and Nebraska Man, a link proposed on the basis of a single tooth which turned out to have come from an extinct pig.

Both of these pseudo-links were introduced as evidence for the theory of evolution at the Scopes trial in 1925. Even the more respectable finds, such as Zinjanthropus, Homo habilis, and the several postulated ancestors named Australopithecus (including the famous "Lucy"), have been rejected or seriously questioned even by evolutionists as genuine "missing links." One of the most troubling aspects of evolutionary thought has been its racist implications. The logic is simple enough: If humans evolved from simpler, less intelligent primates, then perhaps some of are more "evolved" than others. Such racist thinking has accompanied evolutionism from the very beginning, starting with Darwin himself. Darwin visited the South American tribe of the Tierra del Feugians on his journeys and commented that "the difference between a Tierra del Feugian and a European is greater than the difference between a Tierra del Feugian and a beast." Eventually, Christian missionaries discovered otherwise, living among the Feugians and documenting their rich culture and language. Evolutionists may complain that such thinking is not essential to evolution or universal among evolutionists. True enough; but evolutionists cannot make a convincing, rational case against such inferences. Although creationists have themselves not been immune from racism, it turns out that creationism is inherently antiracist while evolutionism offers no protection from racism and can reasonably be construed in its support. The same subjective reasoning that has made it difficult for evolutionists to agree on whether a set of bones comes from a human ancestor, a pre-human "missing link" ancestor, or a distant primate cousin, allows those educated in the evolutionary world view to regard human beings of other races as equal or inferior according to their own predisposed judgments. With all of this in mind, I could not be a Darwinist even if I were not a theist; even more so, as a theist, there is absolutely no reason to attempt to link "theism" or "evolution." Both terms are completely contradictory.[143] Darwinism is in serious trouble throughout secular academia, nor is it compatible with biblical Christianity, so why any church would want to adopt such a contradictory view at theistic evolution, is quite dumbfounding.

I had the privilege of meeting and talking with DR. JOHNATHEN WELLS OF DISCOVERY INSTITUTE in Seattle in the Summer of 2009. Dr. Wells has PhDs from both Yale and Berkeley Universities, with one being in cell and molecular biology. It is interesting that he too confirms the complete fraud of Darwinism. Like me, he is dumbfounded why so many adhere to a theory that has developed into an ideology. In his book, Icon of Evolution, he goes through a list of repeated hoax and outright lies that have been used to support evolution in text books, magazines, museums, and so on for the last century. In regards to human evolution he states with reference to Henry Gee, Chief Science writer for Nature:

> *"The conventional picture of human evolution as lines of ancestry and descent is a 'completely human invention created after the fact, shaped to*

accord with human prejudices.' Putting it even more bluntly, Gee concludes: 'To take a line of fossils and claim that they represent a lineage is not a scientific hypothesis that can be tested, but an assertion that carries the same validity as a bedtime story – amusing, perhaps even instructive, but not scientific.'" [142]

When asked further about this type of information being given to the general public Dr. Wells states:

"The general public is rarely informed of the deep-seated uncertainty about human origins that is reflected in these statements by scientific experts. Instead, we are simply fed the latest version of somebody's theory, without being told that paleoanthropologists themselves cannot agree over it. And typically, the theory is illustrated with fanciful drawings of cave men, or human actors wearing heavy makeup. Whether the ultimate icon is presented in the form of a picture or a narrative, it is old-fashioned materialistic philosophy disguised as modern empirical science." [142]

"Why is it so important to challenge Darwinism? The problem isn't just that Darwinism is false – lots of things are false that nobody worries about. The problem is that Darwinism is no longer merely a scientific theory, but an ideology. An ideology is an all encompassing worldview that attempts to explain everything, often on the basis of a single principle (such as natural selection). Moreover, it demands complete obedience from our hearts and minds."[84]

Nazism, Communism, Fascism, (and as we discussed in the last chapter), Islam are ideologies, in this same sense, Darwinism has also became an ideology to fear.

How does one differentiate between a design and an occurrence?

A design is a pattern of events arranged **with intent** for a purpose. Something that is "designed and made" is created. A creation is the end product of a design – an intention. Intention derives only from a mind. Minds process information, decide upon ends and then adapt means to achieve those predetermined ends. A design reflects a choice made by a mind to affect the future in a particular way.

Let's look at just a handful of other questions that reflect a real problem that has yet to be answered by Darwinists, (if you don't believe me, ask them yourself, (and don't settle for a simple "just because..." or "It is very complicated..."))

> **Irreducible complexity** (IC) is an argument by proponents of intelligent design that certain biological systems are too complex to have evolved from simpler, or "less complete" predecessors, through natural selection acting upon a series of advantageous naturally occurring chance mutations. All this really states is at a molecular level, Darwinism is suppose to tell us how every component of the cell for example was able to evolve into what we see today. For example some of these parts must have all the

part in place to operate (much like a mousetrap), so did they evolve all at once? I will leave it at that for now, but the best answer we can get from Darwinists is that they borrow parts. (the bacteria flagellum, for example, has over 30 parts, and although only around 11 are currently present, Darwinists suggest that they simply borrowed these parts, (not sure where the other 19 came from, but my guess is "chance?")[86]

Specified complexity is an argument proposed by William Dembski and used by him and others to better explain intelligent design. According to Dembski, the concept is intended to formalize a property that singles out patterns that are both specified and complex. Dembski states that specified complexity is a reliable marker of design by an intelligent agent, a central tenet to intelligent design which Dembski argues for in opposition to modern evolutionary theory.

In Dembski's terminology, a specified pattern is one that admits short descriptions, whereas a complex pattern is one that is unlikely to occur by chance. Dembski argues that it is impossible for specified complexity to exist in patterns displayed by configurations formed by unguided processes. Therefore, Dembski argues, the fact that specified complex patterns can be found in living things indicates some kind of guidance in their formation, which is indicative of intelligence. Dembski further argues that one can rigorously show the inability of evolutionary algorithms to select or generate configurations of high specified complexity.[87]

For Dembski, specified complexity is a property which can be observed in living things. However, whereas Orgel used the term for that which, in Darwinian theory, is understood to be created through evolution, Dembski uses it for that which he says cannot be created through "undirected" evolution—and concludes that it allows one to infer intelligent design. While Orgel employed the concept in a qualitative way, Dembski's use is intended to be quantitative. Dembski's use of the concept dates to his 1998 monograph *The Design Inference*. Specified complexity is fundamental to his approach to intelligent design, and each of his subsequent books has also dealt significantly with the concept. He has stated that, in his opinion, "if there is a way to detect design, specified complexity is it."[88]

Dembski asserts that specified complexity is present in a configuration when it can be described by a pattern that displays a large amount of independently specified information and is also complex, which he defines as having a low probability of occurrence. He provides the following examples to demonstrate the concept: "A single letter of the alphabet is specified without being complex. A long sentence of random letters is complex without being specified. A Shakespearean sonnet is both complex and specified."[89]

In his earlier papers Dembski defined *complex specified information* (CSI) as being present in a specified event whose probability did not exceed 1 in 10^{150}, which he calls the universal probability bound. In that context, "specified" meant what in later work he called "pre-specified", that is specified before any information about the outcome is known. The value of the universal probability bound corresponds to the inverse of the upper limit of "the total number of [possible] specified events throughout cosmic history," as calculated by Dembski. Anything below this bound has CSI. The terms "specified complexity" and "complex specified information" are used interchangeably. In more recent papers Dembski has redefined the universal probability bound, with reference to another number, corresponding to the total number of bit operations that could possibly have been performed in the entire history of the universe.[90]

Dembski asserts that CSI exists in numerous features of living things, such as DNA and other functional biological molecules, and argues that it cannot be generated by the only known natural mechanisms of physical law and chance, or by their combination. He argues that this is so because laws can only shift around or lose information, but do not produce it, and chance can produce complex unspecified information, or unspecified complex information, but not CSI. Moreover, he claims that CSI is holistic, with the whole being greater than the sum of the parts, and that this decisively eliminates Darwinian evolution as a possible means of its creation. Dembski maintains that by process of elimination, CSI is best explained as being due to intelligence, and is therefore a reliable indicator of design.

In intelligent design literature, an intelligent agent is one that chooses between different possibilities and has, by supernatural means and methods, caused life to arise. Specified complexity is what Dembski terms an "explanatory filter" which can recognize design by detecting complex specified information (CSI). The filter is based on the premise that the categories of regularity, chance, and design are, according to Dembski, "mutually exclusive and exhaustive." Complex specified information detects design because it detects what characterizes intelligent agency.[91]

So what does this mean in a nutshell? For example, if we stumble across a child dropping letters on the ground, and that they somehow line up to form every word of Shakespeare's Hamlet, we can either infer that they have specified complexity and are therefore a product of design or that their falling into place by mere chance is a better hypothesis.

Where does "Information" come from?

Have you ever asked yourself where "information" itself comes from, or what its source is? "Source" means the origin of something. An **information source** is a source of information for somebody, i.e., anything that might inform a person about something or provide knowledge to somebody. Information sources may be observations, people, speeches, documents, pictures, organizations etc. They may be primary sources, secondary sources, tertiary sources and so on.

Different epistemologies have different views regarding the importance of different kinds of information sources. Empiricism regards sense data as the ultimate information sources, while other epistemologies have different views. So what is the "information source" found in all things; including the cell?

Signature in the Cell

Author and Stephen Meyer at Discovery Institute in August 2009

On June 23, 2009, Harper-One released Stephen C. Meyer's book, "Signature in the Cell: DNA and the Evidence for Intelligent Design". According to the publisher, "Signature in the Cell is the first book to make a comprehensive case for intelligent design based upon DNA. Clearly defining what ID is and is not, Meyer shows that the argument for intelligent design is not based on ignorance or "giving up on science," but instead upon our growing scientific knowledge of the information stored in the cell."[92] The book has been well received by theologians and Intelligent Design proponents[94], as well as garnering a positive review from atheist philosopher Thomas Nagel, who submitted the book as his contribution to the "2009 Books of the Year" supplement for The London Times[95] generating negative responses from within the philosophical community.

In his review for The Times, Thomas Nagel states that "Signature in the Cell...is a detailed account of the problem of how life came into existence from lifeless matter – something that had to happen before the process of biological evolution could begin...Meyer is a Christian, but atheists, and theists who believe God never intervenes in the natural world, will be instructed by his careful presentation of this fiendishly difficult problem."[95] "Signature in the Cell" also made the list of "Top Ten Best Selling Science Books of 2009" at Amazon.com.[96] As of November 2009 the book was entering its fifth printing.[97]

Could one not formulate the following:

1. Information systems come from a mind that designed them.
2. DNA is the product of a massive information system.
3. Therefore, DNA is from a mind that designed the information system.

(While many will say: "no!" I must respectively ask them to then refute these three premises and provide a better hypothesis for where/how information came to be. If this is not accurate, then this should be quite easy for them to do.)

I had the privilege of meeting Dr. Meyer last summer, and he brings up a great question that once again has not been answered by naturalists; where does information come from? Again, we see what Jesus points out that no matter what they see, they won't believe; the same points that Pastor Wurmbrand made to his communist captors. No matter what proof is given; it is ultimately not a matter of evidences, but a matter of acknowledging where the evidences lead. The most ardent atheist now admits that the DNA code of information would fill over 1,000 volumes of Encyclopedia in a very specified set of detailed instructions for life, but yet again, they say it too, is due to mere "chance..."

Give "Chance" a Chance!

CHANCE OF THE GAPS - The fervent belief that religion must be prevented from contaminating science is, as we said, a kind of religious belief itself. One historical factor that has encouraged this belief is the fact that in the past those in the Christian West too easily attributed various features of the natural world to direct supernatural agency, only to have some scientist come along and demonstrate a regular natural phenomenon to be at work. But to swing the pendulum to the other extreme and disallow the activity of God as a possible explanation for anything, regardless of the evidence, is also unwarranted. Both of these extremes – uncritical supernaturalism and uncritical naturalism — should therefore be avoided. The attribution of unexplained phenomena (planetary orbits, meteors, earthquakes, volcanoes, and the like) to supernatural intervention by God has often been criticized as a "God of the gaps" approach. But just as irrational is the assumption made by many naturalists that God never intervenes in his creation and that everything, even the very existence of the universe, must be explainable in natural terms — what has been called a "Nature of the gaps" approach.36 Similarly, Hugh Ross has criticized the appeal by cosmologists to the chance fluctuations posited by quantum theory to explain the origin of the universe as a kind of "Chance of the gaps" methodology.[98]

Design encourages scientists to look for function where evolution discourages it. Or consider vestigial organs that later are found to have a function after all. Evolutionary biology texts often cite the human coccyx as a "vestigial structure" that hearkens back to vertebrate ancestors with tails. Yet if one looks at a recent edition of Gray's Anatomy, one finds that the coccyx is a crucial point of contact with muscles that attach to the pelvic floor. The phrase "vestigial structure" often merely cloaks our current lack of knowledge about function. The human appendix, formerly thought to be vestigial, is now known to be a functioning component of the immune system. Admitting design into science can only enrich the scientific enterprise. All the tried and true tools of science will remain intact. But design adds a new tool to the scientist's explanatory tool chest. Moreover, design raises a whole new set of research questions. Once we know that something is designed, we will want to know how it was produced, to what extent the design is optimal, and what its purpose is. Note that we can detect design without knowing what something was designed for. There is a room at the Smithsonian filled with objects that are obviously designed but whose specific purpose anthropologists do not understand.[99]

So it is really up to you now. Is the brilliant theory of "chance" for everything sufficient? Or does the theory that there is an intelligence starting to make more sense? I have only touched on a few very basic points, but this should be enough, Lord willing, to at least establish Deism, (that there is some type of creator or designer behind all of life).

So what is the next step?

Now based on the brief information above, we have established that the hypothesis of a Designer, Creator, God, etc, is not only the strongest hypothesis, but this would in theory explain the origins of space, time, information, consciousness, meaning, purpose, morals, and so on, while all naturalistic theories do not. (And as we have established, naturalistic theories rule out a naturalistic origin; the scientific law of causality, biogenesis, and probability alone point solely to a Designer, (while naturalists/Darwinists attempt to say, (while this is accurate), these laws of science do not apply to the singularity of the big bang, the first life, etc. (it must be nice to simply pick and choose when and where you would like to use established scientific principles and criteria.))

Moreover, the fact remains that there being some type of supreme "Creator" throughout the history of humanity, and today is believed by over five billion people groups. Why? Once again, the God hypothesis answers this simply because God's existence has left this image, or feeling in each one of us. Naturalism must simply say this is another accident/coincidence, that we have evolved to the point that our brains somehow, throughout history, have just happened to evolve into our brains. So once again, the God hypothesis fills this answer while naturalism/Darwinism seems to be a much weaker and less satisfactory hypothesis.

(Much more could be explored on this subject that ties in nicely with a biblical account; just as Newsweek magazine's January 11th 1988 article (pp. 46-52) "The Search For Adam and Eve" where microbiologists stated they had looked at assortments of genes and DNA that led them to a single woman from whom we are all descended. I will not dive into this anymore in great detail here, but will just mention these facts that while the scientists usually attempt to distance themselves from the Bible, they can not dismiss the fact that these evidences continually seem to make great sense when contrasted to the accounts given in the Bible of the Old and New Testaments. Another few examples could be the fact that most scientists and anthropologists agree that the world did indeed have a single language at one time (see Tower of Babel), that when we look at geological evidences, the facts that there are over 200 flood legends closely resembling that of the Noah narrative found in the Bible, and that our first civilizations did come from the same area near Mesopotamia that the Bible records; and recently, that geologists found a huge body of water under the earth's crust the size of an ocean; all give at least a moderate cumulative case argument that someone arguing for Noah's flood has just as much credibility as an atheist shouting "mere coincidences!" All I am saying, is that once we start looking at all the facts, we can make a cumulative case argument that all of this evidence does indeed point to a designer or grand architect, and that of the evidences studied so far (whether coincidence or not), the biblical account falls right in line with all mainstream fields of study; but for right now, we will just leave the evidences as secular, and not dive into the coincidences of them matching a biblical account.)

At this point everyone should at least acknowledge that the scientific evidences and information point us strongly to Deism, (that there is a God), but not necessarily Theism (that it is a personal God that is still involved in our lives today and in the future). If a person denies this, then they must satisfactorily refute our above points, and show that they have a stronger hypothesis for each of those, (I don't accept "chance" as a logical answer that has more explanatory power, than that of intelligent design, and more and more scientists are agreeing with me.)

So a person might say at this point: "Okay – there is obviously some type of supreme architect that put everything together; but why theism, and moreover, why Christ?"

Let us first look at three principles that a truth about God should be able to address:

1. Logical consistency
2. Empirical Adequacies
3. Experiential Relevance

Once again, while there are a ton of additional proofs, I will keep it simple and only address a few that fill these criteria, starting with our moral comprehension.

Have you ever asked yourself why something is right or wrong, (good or bad)?

Once again, there are many books on this subject alone, but I will give you the basic summaries for the skeptic to ponder, and for the Christian theist to hopefully grow even more confidence in the faith that they profess, as well as the confidence to share the message more boldly, with gentleness and respect.

The Moral Argument – Are Morals Objection or Subjective?

Richard Wurmbrand provides a vivid example of morality during his 14 year prison sentence in communist prison: "The cruelty of atheism is hard to believe. When a man has no faith in the reward of god or the punishment of evil, there is no reason to be human. There is no restraint from the depths of evil that is in man. The Communist torturers often said, 'There is no God, no hereafter, no punishment for evil. We can do what we wish.' I heard one torturer say, 'I thank God, in whom I don't believe, that I have lived to this hour when I can express all the evil in my heart.' He expressed it in unbelievable brutality and torture inflicted on prisoners." (Tortured for Christ, pg 36)

When I say objective, this means that regardless of what someone says, a certain act or action is truly right or wrong, regardless of geography, culture, tradition, or time. On the other hand, subjective would be that I prefer vanilla ice cream over chocolate. This does not mean vanilla is better tasting than chocolate for everyone, in all places and at all times. It is simply my preference. So with that being said, are moral values objective or subjective?

Is rape, torture, or murder simply subjective? In other words, while we might not like it, there is nothing actually wrong about it, it is simply our preference against it, (much like the flavor of ice cream analogy). Or are such acts truly, objectively wrong?

From a Darwinist approach, our minds and society have just accidentally by "chance" evolved in such a way that our societal instincts have deemed such acts as torture, rape, child abuse as wrong, but it isn't really wrong. In other words, if we had evolved a little differently, then such acts might be deemed acceptable.

So the Darwinist/Naturalist once again can not satisfactorily answer this concern, for they must either say that nothing is wrong with such horrible acts, and that there is no real difference between a Hitler or Mother Teresa, but we just happened to by chance, evolve this way. While most will agree this is not a satisfactory conclusion, we must acknowledge if we say that there are truly some things that are right/wrong or good/evil, then we say that moral values do have an objective meaning. But if we say this, how is it objective? Just like the information in DNA, this objectiveness of morals had to come from some type of moral law giver. For how can morals be objective, and not have derived from some type of creator/designer that decided what was right/wrong or good/evil in the first place?

Let me use another example – If the Nazis had won WW2 and executed everyone that said the genocide of the Jews was wrong, (so that everyone in the world said that the genocide was good); would this act truly be "good" then? From the Darwinist/Naturalist perspective, the answer would be "yes," because society through natural selection, had changed to say that the killing and torture of a Jewish child is a good act. I believe (and hope) that we all can see that such an act would still be "wrong/evil" even if no one on earth condemned it. It is wrong, because the consciousness built into us, cries out that such an act is objectively wrong.

So while you are thinking about this, let me make the following observation:

1) **If God does not exist, then objective moral values do not exist.**
2) **Objective moral values do exist.**
3) **Therefore, God exists**

I find this a very elementary step that helps bridge the gap between deism and theism, but not completely. I have had some say there are objective moral values, but that does not necessarily point to a God/Law giver. But I have yet to ever receive an answer as to how. Every time someone tries to explain these morals outside of a law giver, they describe them in the tone of being subjective, not objective. Some may still argue that it is therefore not wrong to kill, rape, or torture, but unless they are indeed a sociopath, they do not truly believe this, (and they never live it); therefore the best hypothesis over "chance occurrence" of morals, is that a Creator/Designer programmed in us that such acts as genocide are objectively wrong, and acts such as love are objectively good. I will just leave you with this simple question to ponder: How can anything be right or wrong objectively, outside of God?

While much more could be spent on this subject, I think enough has been said, and we will therefore dive briefly into the reasons that further bridge Christ to this deism/theism.

> *"Despite everything, I believe that people are really good at heart."*
> Anne Frank

Don't we decide what truth is for ourselves?

I suppose at this point, I should briefly touch on the fact of other religions, so that one will have a proper understanding of the pluralism/relativism that is so prevalent amongst young adults today. Again, while there are complete books on just these two subjects, I will only touch on them logically, (and direct you to further research on your own, or reference you to the recommended reading list at the back of the book).

Long story short, relativism states that truth is relative to the individual. In other words, if I say a blueberry is the color blue, and you say it is green, then we are both accurate, because each statement is true to that person. But surely this does not make sense? So if I say the earth is flat and you say the earth is round, does that mean that the earth "truly" is flat? Of course it does not. The earth is round no matter who says otherwise. If a medicine bottle says "poison," is it relatively true, or will it kill both individuals regardless of what they think? Obviously, the ingredients in the poison will kill regardless of what they individually "believe." So I honestly find it quite dumbfounding that relativism has permeated society and schools as much as it has. If one wishes to believe this, they will only in words but not in actions. (when someone takes poison and says "I believe it not to be poison," and therefore they come away unhurt, then I will look at relativism as a legitimate option, but until then I will waste no further words on it). *Just ask your "relativist" friend if the earth is flat because you or they think it is and see their response.

Pluralism is an equally illogical mindset that through "political correctness," has begun to be espoused via media and on our school campuses. It too requires you to "check your brain at the door," because the vast majority of person either do not understand what it means, or they say (though it is not logical), it is a politically correct/tolerant mindset, and therefore it is acceptable over truth. Pluralism basically states that whatever path you choose, will get you to the same destination. Religious pluralism therefore means that whether it is a Buddhist, Muslim, Hindu, Christian, or any of the other thousands of smaller religions, all lead to "God," and therefore, all lead to the same being. While this does sound nice, we must dig a little deeper before we make such claims.

-Traditional Buddhism states for example, that there is no actual "God," but that enlightenment is the ultimate goal, (there is not good or bad because it is all illusory).

-Hinduism follows more of the route of pantheism, that all is God, (while other branches hold that there are thousands of Gods).

-Judaism states there is one God.

Obviously there are some contradictory terms here, so if we follow that there is only one God, there is no God, all is God; I hope that we can all agree that these terms are very contradictory, so to simply state that "there are many roads, but they all lead the same place;" one must at least understand what this fully means. This is all the time I feel is warranted to spend on these two subjects, but I wanted to at least clear the air for those who from the start will simply say, "all is relative or pluralistic."

"Moral neutrality seems virtuous, but there's no benefit, only danger. In our culture we don't stop at 'sharing wisdom, giving reasons for believing as we do – and then trusting others to think for themselves,' nor should we. This leads to anarchy. Instead, we use moral reasoning, public advocacy, and legislation to encourage virtue and discourage dangerous and morally inappropriate behavior.

That is, if we haven't been struck morally paralyzed by relativism."[165]
(Koukl/Beckwith – Relativism – Feet Firmly Planted in Mid-Air)

Why Jesus Christ?

After World War 2 ended, Chancellor Konrad Adenauer of West Germany and mayor of Cologne, upon looking out his office window at the destruction caused from the war; asked Billy Graham if he really believed in the Resurrection of Jesus Christ. Billy Graham somewhat confused by the question said, *"Yes, I do. Christianity wouldn't be Christianity without it."* Adenauer then said, *"So do I, and if Jesus Christ was not raised from the dead, there's no hope for the human race."* [147]

Having done my undergraduate work in History, I have always found it quite alluring that Judaism and Christianity are the only religions that use history to substantiate their claims. Islam, for example, follows that the Qur'an describes itself as a book of guidance, rarely offering detailed accounts of specific historical events, and often emphasizing the moral significance of an event over its narrative sequence, but it is a revelation of ONE witness: that of Mohammad. (To even ask the question of which is more relevant: a text on Jesus Christ and the New Testament written 600 years earlier than the Koran by a multitude of authors and first-hand witnesses, or by the interpretation of ONE individual 600 years later; is to answer your own question).

Buddhist texts are more on the life and teachings of Buddha and his teachings.

Hinduism regards the Vedas as a collection of hymns or mantras to be chanted by a priest; Atharvaveda as a collection of spells and incantations.

When we really dive into all other religions and belief systems and their writings, we find the fact that the Bible is the only religious documentation that is steeped in:

1) **Creation/origin science documentation**
2) **Historic reliability**
3) **Prophetic texts**

This means that it can be followed/examined through a historic approach and thus testable, (at least through our limited means of evaluation). Therefore, there are ample reasons to spend a few moments in evaluating the historicity and reliability of this text; Christ Himself used the Old Testament to point to His fulfillment of these texts, so it is important to quickly evaluate, (and be able to explain to others), these three points, to see if they hold up to the test.

As I have continually repeated, we will only review a brief summary of these main points, and reference you to much more in depth and knowledgeable authors and their works at the end of this book.

One of the first things we must establish is that when we say manuscripts, we are using the term for historic writings that were used in piecing together what we now simply call the Bible. Some will argue that because a text was incorporated into the Bible, that it now can not be relied upon to be unbiased, but this is simply fallacious. Scribes copied down these manuscripts in various places, languages, and styles, and then through a painstaking evaluation of these texts, the text of the Bible slowly began to be put together. Therefore these manuscripts existed independently, long before the Bible was put together.

We have already covered premise one, on the creation story at the first of this chapter. The universe did indeed have a beginning. Looking at a neutral and credible source of reference to the Bible's creation story, noted astronomer and agnostic Robert Jastrow notes in the first chapter of his book "God and the Astronomers" that "When an astronomer writes about God, his colleagues assume he is either over the hill or going bonkers. In my case it should be understood from the start that I am an agnostic in religious matters."[100]

Jastrow then concludes that his studies in the beginnings of the universe match up remarkably well with the biblical account of creation: *"Now we see how the astronomical evidence leads us to a biblical view of the origin of the world. The details differ, but the essential elements in the astronomical and biblical accounts of Genesis are the same: the chain of events leading to man commenced suddenly and sharply at a definite moment in time, in a flash of light and energy."* [100]

Many believers and unbelievers alike have commented that the biblical account of creation match precisely with what they observe in their scientific studies.

So the Biblical account holds up well with step 1, so let's review step 2.

While nowhere in the Bible is there a promise of purity of text throughout history, there is a great deal of evidence that suggests that the Bibles we read are extremely close to the original. Such reliability helps support the consensus of the Bible being a valuable and accurate account of history as well as revelation from God. We must remember that the Old and New Testaments cover a time span that we can verify with archaeology and other sources, from approximately 2,200 BC (Abraham) to the Book of Revelation, approximately 90 AD; so to keep a faithful written account would indeed prove challenging. Looking first at the Old Testament, we find that the accuracy of the copies we have are supported by a number of evidence. All of the copies we have agree with the majority of text, (while some seem to be more paraphrased than a word for word account). These also agree with our more modern copies such as the Septuagint (the Greek translation), which dates from the 2nd century AD. Finally with the discovery of the Dead Sea Scrolls, we are able to compare all of our documents to these scrolls that date from approximately 250 BC, and provide an almost complete account of the Old Testament. When we compare all of these, we find that all of these translations are extremely close, and furthermore are extremely close to our modern translations. Most scholars have agreed that these match our current translations by 95% with the 5% mostly being variations in spelling that do not alter any themes. We also have ancient documents quoting the Old Testament that also are shown to act as further confidence of the reliability of the Old Testament.[101]

ARCHAEOLOGY –

Another interesting field of study is that of Biblical Archaeology. While I won't get into the details, historians agree that archaeology has consistently and repeatedly confirmed the history of the Old Testament. Many history books have been re-written after finding archaeological discoveries that proved the biblical story accurate, but was not at the time, considered accurate by the historian. It is ironic, that Biblical Archaeology really got started about 150 years ago, and its purpose was more for disproving the Bible, rather than proving it. But once again, just like scientists and historians, the more archaeologists attempted to disprove the biblical narratives; the more they ended up confirming their truth. Thus began a fascinating field of study that not only gives further confirmation of Biblical truths, but also gives us a real insight as to how these people lived. It is a shame that this study only came about in the last 150 years, or we would even have more history at our disposal.

One fascinating book that first reviewed the correlation of Old Testament archaeology with that of history, was *"The Bible as History"* by Dr. Werner Keller whose book sold more than 10 million copies. In it, Dr. Keller traces the Old Testament and New Testament histories through that of confirmed archaeological finds. Little did Dr. Keller know that the next 40 years would have a large amount of finds that would further confirm this history. After many of these confirmations, Dr. Keller commented:

> *"These breathtaking discoveries, whose significance it is impossible to grasp all at once, make it necessary for us to revise our views about the Bible. Many events which previously passed for 'pious tales' must now be judged to be historical. Often the results of investigation correspond in detail with the biblical narratives. They do not only confirm them, but also illumine the historical situations out of which the Old Testament and the Gospels grew."* [102]

More recently Dr. J. Randall Price (Th.M. Old Testament and Semitic Languages and Ph.D., Middle Eastern Studies), wrote a book called *The Stones Cry Out (1997)*, in which he goes through the basics of biblical archaeology and what it reveals to us about scripture. While remaining humble on archeology's role, he says:

> *"According to Webster's English Dictionary, one of the meanings of the word confirm is 'to give new assurance of the validity' of something. Archaeology provides a new assurance of the Bible from the stones to accompany the assurance we already have from the Spirit. The value is an apologetic one, and from the beginning of the science of archaeology, it was a contributing factor in both instigating and sponsoring excavations. Almost all scholars still attest to the significant agreement between the stones and the Scriptures."* [103]

Once again, I am not going into a lot of details in this booklet, I will just simply say that while questions always remain, we can conclude that archaeology has confirmed the history of the Old Testament, not only in its general outline, but in many of the minute details also. And as for the New Testament, after the Old Testament period of the Judges, the archeological evidence becomes increasingly clear that the biblical authors knew very well what they were talking about.[104]

NEW TESTAMENT

As we enter into the era of the New Testament, the facts become quite clear and in all honesty overwhelming. When it comes to the New Testament, no other ancient writing is even close to being as well attested or documented as the New Testament writings. For example, we have to date at total of 5,664 Greek manuscripts of the New Testament; when we add the Latin, Ethiopic, Slavic, Armenian and others, the total exceeds 24,000! Moreover, when we look at the manuscript evidence of other writings such as Cesar's Gallic Wars we have a total of 10 manuscripts to support it; Herodotus' History has 8, Tacitus Annals has 20, and Homer's Iliad has an impressive 643. These are not even close to the reliability we have in the New Testament. But it doesn't stop there; Cesar's Gallic Wars were written in 100 BC, but the earliest copy we have is dated 900 AD, Herodotus' History was written 400 BC and our earliest copy is dated 1300 AD, Tacitus Annals was written in 100AD and our earliest copy is dated 1100 AD.[105]

So what about the New Testament? Most historians and critics alike agree that the Gospels were written between 45 – 90 AD, (with Mark being first and John last), and that the Pauline Epistles can be dated within 10-15 years of Christ's crucifixion, (some even estimate the events to be less than five years after Christ's crucifixion). We must also remember these were written by eye witnesses, and multiple times attested, unlike most other religious texts, (for example, no one saw the Buddha have his visions, nor did anyone other than Mohammed witness his revelations from the angel Gabriel). What we have is the most ancient document attested writings in history, the fact that we have multiple first-hand accounts from Jesus' followers, and all were written shortly after Jesus' death and resurrection. As Ravi Zacharias states:

> *"In real terms the New Testament is easily the best-attested ancient writing in terms of the sheer number of documents, the time span between the events and the document, and the variety of documents available to sustain or contradict it. There is nothing in ancient manuscript evidence to match such textual integrity..."* [106]

We must remember that I am just scratching the very surface of our evidential criteria; as the brilliant Dr. John Warwick Montgomery also points out, not only do we have the remarkable internal evidence, but we also have a plethora of external evidence. For example, while 11 of the 12 apostles were martyred for their faith, John (who walked with Jesus, witnessed his crucifixion first hand, witnessed his resurrection first hand, and took care of Mary the Mother of Jesus until her death; also had first hand students of his own, who wrote down what he said and did. Of these were Papias, bishop of Hierapolis around AD 130, Polycarp, bishop of Smyrna, and others were students who walked with John first hand. We then have Irenaeus who was a student of Polycarp, and his writings also, match what we are told in the Gospels, and so it goes down through history. [107]

Historians and literary critiques continue to be amazed by the level of accuracy of the scriptures. "Classical scholar and historian Colin Hemer chronicles Luke's accuracy in the book of Acts verse by verse. With painstaking detail, Hemer identifies 84 facts in the last 16 years of Acts that have been confirmed by historical and archaeological research.[108,109] Roman historian A.N. Sherwin-White says, "For Acts the confirmation of historicity is overwhelming...Any attempt to reject its basic historicity must now appear absurd. Roman historians have long taken it for granted."[110] We also find another 59 historically confirmed details in the Gospel of John, which when you add this with John's personal conversations with Jesus, it really does seem that it would take more faith to dismiss these than to take them as authentic. When we look merely at the three books of Acts, John, and Luke, we find 140 historically confirmed details, plus the fact that they continually reference historic figures of that time, we can see how this overwhelming adds further credit to the New Testament's reliability and historicity, as well as what the Apostles themselves had to say.

Aside from the over 24,000 manuscripts that match to a degree of 99.9%, we also have alternative sources that further collaborate the basics outline of the Gospels.

NON BIBLICAL SOURCES

Tacitus (A.D. 56 – ca. 117)- First-century historian; Tacitus is considered one of the most accurate historians of the ancient world, wrote: *"Consequently, to get rid of the report, Nero fastened the guilt and inflicted the most exquisite tortures on a class hated for their abominations, called Christians by the populace. Christus, from whom the name had its origin, suffered the extreme penalty during the reign of Tiberius at the hands of one of our procurators, Pontius Pilatus, and a most mischievous superstition, thus checked for the moment, again broke out not only in Judea, the first source of the evil, but even in Rome, where all things hideous and shameful from every part of the world find their center* *and become popular." (Tacitus, A, 15.44)* -The "mischievous superstition" to which Tacitus refers is most likely the resurrection of Jesus. The same is true for one of the references of Suetonius which follows.

Josephus (A.D. 37 – sometime after 100) - Was a Pharisee of the priestly line and a Jewish historian working under Roman authority; he was a Jew and not a follower of Christ, but in his *"Antiquities of the Jews, SVIII, 33)*, had this brief description of a man called Jesus: *"Now there was about this time Jesus, a wise man, if it be lawful to call him a man, for he was a doer of wonderful works, a teacher of such men as receive the truth with pleasure. He drew over to him both many of the Jews, and many of the Gentiles. He was the Christ, and when Pilate, at the suggestion of the principal men among us, had condemned him to the cross, those that loved him at the first did not forsake him; for he appeared to them alive again the third day; as the divine prophets had foretold these and ten thousand other wonderful things concerning him. And the tribe of Christian so named from him are not extinct at this day."* -This from a non-Christian historian... (Josephus also confirmed the existence of John the Baptist: *"Now, some of the Jews thought that the destruction of Herod's army came from God, and very justly, as a punishment of what he did against John, who was called the Baptist; for Herod slew him, who was a good man, and commanded the Jews to exercise virtue, both as to righteousness towards one another and piety towards God, and so to come to baptism." (Josephus, AJ, 18.5.2)* Josephus also refers to the martyrdom of James the brother of Jesus.

Suetonius - Was chief secretary to Emperor Hadrian, (who reigned AD 117-138); he confirms the report recorded in the Book of Acts 18:2, that Claudius commanded all Jews to leave Rome in AD 49, (Life of Claudius, 25.4). Speaking of the aftermath of the great fire at Rome, Suetonius reports, *"Punishment was inflicted on the Christians, a body of* *people addicted to a novel and mischievous superstition."* *(Life of Nero, 16)*

Thallus - Wrote around AD 52; was quoted in reference to the darkness that followed the crucifixion of Christ: *"On the whole world there pressed a most fearful darkness, and the rocks were rent by an earthquake, and many places in Judea and other districts were thrown down. This darkness Thallus, in the third book of his History, calls, as appears to me without reason, an eclipse of the sun." (Chronography, 18.1)* -Described also in Luke 23:44-45.

Lucian - 2nd Century Greek writer - *"The Christians, you know, worship a man to this day—the distinguished personage who introduced their novel rites, and was crucified on that account... You see, these misguided creatures start with the general conviction that they are immortal for all time, which explains the contempt of death and voluntary self-devotion which are so common among them; and then it was impressed on them by their original lawgiver that they are all brothers, from the moment that they are converted, and deny the gods of Greece, and worship the crucified sage, and live after his laws. All this they take quite on faith, with the result that they despise all worldly goods alike, regarding them merely as common property." (Lucian of Samosata, DP, 11-13)*

Emperor Trajan - The emperor gave the following guidelines for punishing Christians: *"No search should be made for these people, when they are denounced and found guilty they must be punished, with the restriction; however, that when the party denies himself to be Christian, and shall give proof that hi is not, (that is, by adoring our gods), he shall be pardoned on the ground of repentance even though he may have formerly incurred suspicion."* (Pliny the Younger, L, 10:97)

Pliny the Younger - Roman author and administrator
Mara Bar-Serapion - Syrian writer

As Norm Geisler and Frank Turek point out, these are is just a few "non-Christian" historians and writers that confirm the biblical account of Christ; I encourage you to continue this search if you would like further witnesses and/or proof.

When we only use 100% non-Christian sources of history, we come up with the following facts:

1) Jesus lived during the time of Tiberius Caesar.
2) He lived a virtuous life.
3) He was a wonder-worker.
4) He had a brother named James.
5) He was acclaimed to be the Messiah.
6) He was crucified under Pontius Pilate.
7) He was crucified on the even of the Jewish Passover.
8) Darkness and an earthquake occurred when he died.
9) His disciples believed he rose from the dead.
10) His disciples were willing to die for their belief.
11) Christianity spread rapidly as far as Rome.
12) His disciples denied the Roman gods and worshipped Jesus.

In light of these references, this is yet another affirmation of the New Testament's accuracy.[111]

So while only touching on a few points, we should be able to confidently say that premise two has been successfully addressed, on the historical reliability of the Bible in which we learn about Jesus, (and which Jesus used Himself: today's Old Testament). No other religion can follow these lines of evidence to the degree of the Bible or Christianity.

Premise 3 – Prophecy is quite unique to the writings of the Bible. Both Old and New Testament have numerous accounts of prophetic messages that are fulfilled. You can say as the naturalist does in most things, it is coincidence, chance, or luck, or you can take a more logical approach and begin researching the probabilities of the first two points, and see how this begins painting an overwhelming picture of a huge puzzle called "purpose/life/existence" etc, and see how they are indeed beginning to form a beautifully complete picture, that no other worldview can.

Instead of giving you all of the different prophecies foretelling the coming "Messiah," I will encourage you to research these on your own. From His birth, life, and death, all prophesied by many of the different Old Testament prophets, all lead to the fulfillment of the chosen Son of God, the Messiah. This in itself is quite amazing, that through many different individuals, none contradicted one another in their foretelling of the coming Messiah – all are filled by one person, in Jesus Christ. (Please remember that we have these prophecies recorded in the Dead Sea Scrolls which were written well before Jesus of Nazareth was born) If we take some of these prophecies it is easy to calculate the probability of someone fulfilling such criteria.

The Mathematical Odds of Jesus Fulfilling all Prophecies

The following probabilities are taken from Peter Stoner in Science Speaks (Moody Press, 1963) to show that coincidence is ruled out by the science of probability. Stoner says that by using the modern science of probability in reference to eight prophecies, 'we find that the chance that any man might have lived down to the present time and fulfilled all eight prophecies is 1 in 10 to the 17 power." That would be 1 in 100,000,000,000,000,000. In order to help us comprehend this staggering probability, Stoner illustrates it by supposing that "we take 10^{17} silver dollars and lay them on the face of Texas. They will cover all of the state two feet deep.

Stoner considers 48 prophecies and says, "we find the chance that any one man fulfilled all 48 prophecies to be 1 in 10^{157}, or 1 in

100,000,000,000,000,000,000,000,000,000,000,000,00 0,000,000,000,000,000,000, 000,000,000,000,000,000,000 ,000,000,000,000,000,000,000,000,000,000,000,000, 000, 000,000,000,000,000,000,000,000,000,000,000,000,000.

The estimated number of electrons in the universe is around 10 to 79th. It should be quite evident that Jesus did not fulfill the prophecies by accident."[112]

So why the naturalist or naturalistic historian will still scream "another lucky chance/coincidence," I think Christianity has successfully filled all 3 of these premises. So we can conclude:

1) Any document that can be confirmed to be accurate through creation, historicity, and prophecy would be the most reliable document in existence, and could be fully trusted.
2) The Bible fulfills this criteria
3) Therefore the Bible can be fully trusted.

One may argue with these points, but they must once again, disprove the three points as I have lined them out, or present a more probable hypothesis, (Please encourage them to formulate a better hypothesis than "luck/chance" for each)

So you might ask why then do so many now argue against Christianity? Some are simply due to Christians giving Christ such a bad name, they are disgusted with the entire "religious" thing, (which we'll discuss in the next chapter); most have presuppositions or biases against the miracles discussed in the New Testament or on the resurrection of Christ.

When discussing a non-naturalistic miracle, or the Resurrection, we must remember that scientists are out of their fields on this subject, and we must turn to the historian. For example, science can show that it is possible that the Nazi's could have some through the Ardennes forest to launch a stealth attack against the Allied forces, but it can not prove they did. Therefore, we must look to history. Historic evidences included could be: witnesses, written documentation, dates of written composition, supporting texts from different sources, and archaeology to name just a few.

A great example of this was when Philosopher and Theologian William Lane Craig debated the staunch Darwinist Peter Atkins.[113]

Atkins stated: "Everything in the world can be understood without needing to evoke a God. You have to accept that's one possible view to take about the world."

Craig: "Sure, that's possible, but..."

Atkins: "Do you deny that science can account for everything?"

Craig: "Yes, I do deny that science can account for everything."

Atkins: "So what can't it account for?"

Craig calmly said he would just name five:

1) Mathematics and logic – science can not prove them because science presupposes them.
2) Metaphysical truths – such as, there are minds that exist other than my own.
3) Ethical judgments – you can not prove by science that the Nazis were evil, because morality is not subject to the scientific method.
4) Aesthetic judgments – the beautiful, like the good, can not be scientifically proven.
5) Science itself – the belief that the scientific method discovers truth can not be proven by the scientific method itself.

(Atkins had no immediate response other than a somewhat dumbfounded look on his face.)

With this being said, we must dismiss our presuppositions and prejudices against such things as "miracles," and just see what the best hypothesis surrounding Christ's resurrection is.

We will only touch on a few, to see how credible the resurrection is, and why you should be able to easily share this with a skeptic, (as well as quell any of your own doubts).

A Resurrection???

How do we know from a historical perspective that Jesus rose from the dead? When working on my undergraduate studies in history at the University of Arkansas, I began to see how historians put together their information from past events, to arrive at a working cumulative case argument from the facts at hand. When we take this same approach with an agnostic or skeptic, we have no real choice, (regardless of our preferences), but to take the facts as they are, and see where the supporting facts lead us. So in light of the question concerning a skeptic asking how we know that Jesus rose from the dead, we should approach it the same way we would if someone asked us, "How do you know Alexander the Great never lost a battle?" And that is by taking the most agreed upon historic established facts, and then going down the list of less established facts from there, until we can arrive at a consensus, or best explanation, for our conclusions; and that our conclusions form the best hypothesis, (over competing ones), for the claim. Here are just three of the **established facts**, (which for the most part, are agreed upon by critic and non-critic alike[114]), concerning the facts around Jesus' death:

1) **The empty tomb**
2) **The post mortem appearances**
3) **The origin of the Christian faith**

We will briefly review these three claims – **The Empty Tomb** must have been empty because the disciples could not have believed in Jesus' resurrection if his corpse still lay in the tombs. Also, Jews or Romans could have simply presented the body before the populace to quell any dispute about the missing corpse. One of the most remarkable facts about the early Christian belief in Jesus' resurrection was that it flourished in the very city where Jesus had been publicly crucified. Few would have been prepared to believe such nonsense as that Jesus had been raised from the dead, if they had not had a real reason to justify this belief. I will paraphrase Dr. William Lane Craig[114] in pointing to the fact too, that Jesus' burial is multiply attested in extremely early, independent sources.

The account of Jesus' burial by Joseph of Arimathea is part of Mark's source material; this being a very early source which is probably based on eyewitness testimony and which some critics date to within seven years of Jesus' crucifixion. Moreover, Paul in 1 Corinthians 15:3-5 quotes a much older Christian tradition that he had received from the earliest disciples, (within the first five years of Jesus' death): "that Christ died for our sins in accordance with the Scriptures, and that He was buried, and that He was raised on the third day in accordance with the Scriptures, and that He appeared to Cephas, then to the Twelve." For these and other reasons, most New Testament critics agree that Jesus was buried by Joseph of Arimathea in a tomb, and it is one of the earliest and best-attested facts about Jesus. As mentioned, if this conclusion is correct, then it seems very difficult to deny the historicity of the empty tomb.

The post mortem appearances - Once again looking at the very early letters of 1 Corinthians 15:3-8:

> "For I delivered to you as of first importance what I also received, that Christ died for our sins in accordance with the scriptures, and that he was buried, and that he was raised on the third day in accordance with the scriptures, and that he appeared to Cephas, then to the Twelve. Then he appeared to more than five hundred brethren at one time, most of whom are still alive, though some have fallen asleep. Then he appeared to James, then to

> *all the apostles. Last of all, as to one untimely born, he appeared also to me."*

If we briefly break this down, it tells us that Jesus appeared to Peter, to all of the apostles, to more than 500 brethren, to James, and finally to Paul. If we go straight to the 500, we find something quite amazing. Paul reports that "most are still alive," (which is basically saying) - "they are witnesses too, so don't hesitate to go ask them as well," (which would also help explain why a huge populace of the city believed in the resurrection instead of just a handful of fanatics). Also what is very interesting is James - Jesus' Brother, who by most accounts did not believe in Jesus' Messiahship. After this alleged appearance, James became a large player and instrument in the early church, and as recorded by the Jewish historian Josephus, was stoned to death for his faith in Christ. (Josephus, Antiquities of the Jews) Then finally, He appeared to a well trained Jewish Pharisee, Saul of Tarsus, who was a vehement persecutor of the early Christian church. This is only a very brief sketch, but we are once again building a cumulative case argument that is increasingly convincing, and hard to dispute.

The origin of the Christian Faith - Even the most critical of New Testament scholars holds that the disciples at least did witness what they "thought" was the resurrected Jesus. It is becoming increasingly difficult to admit that the early and rapid growth of the Christian faith could be attributed to anything else. Jewish tradition had always held to a Messiah that would come in triumphantly to establish the throne of David, not one who would shamefully be executed by them as a criminal. It is difficult to imagine what a disaster the crucifixion was for the disciples' faith. Jesus' death on the cross must have looked like a humiliating end for any hopes entertained that he was the Messiah. But the belief in the resurrection of Jesus reversed the catastrophe of the crucifixion, because God had raised Jesus from the dead as he had foretold, he was seen to be the Messiah after all. This is why the majority of all attempts to explain away the resurrection has thus collapsed. The origin of Christianity owes itself to this belief held by the earliest disciples, that God had raised Jesus from the dead. This belief cannot be plausibly accounted for in terms of either Christian, pagan, or Jewish influences. As Dr. Craig points out: "Even if we grant, for the sake of argument, that the tomb was somehow emptied and the disciples saw hallucinations - suppositions which we have seen to be false anyway - the origin of the belief in Jesus' resurrection still cannot be plausibly explained. Such events would have led the disciples to say only that Jesus had been translated into heaven, not resurrected. The origin of the Christian faith is therefore inexplicable unless Jesus really rose from the dead."[114] *(Reasonable Faith, pg 395)* It is the historian's goal, using all his or her critical skills, to determine what happened in the past by reconstructing it on the basis of evidence. As Fredrick Copleston states in "Problems of Objectivity," in On the History of Philosophy:

> *"The historian is not free to interpret the texts as he likes. Some statements may be ambiguous; but there are others, the meaning of which is clearly determined independently of the historian's will. For example, he is not at liberty to deny the fact that Marx asserted the priority of matter to spirit of mind. As far as the historian is concerned, the texts constitute something given, something which limits his reconstruction."*[115]

So whether one agrees with Marxism or not, it is irrelative to the facts. Moreover, it is not up to the historian to rule out the resurrection simply because they feel personally that miracles such as the resurrection do not fit their pre-suppositional beliefs. Therefore, based on the above mentioned, and relatively non-controversial facts, plus the multiple confirmed facts by both first-hand and later witnesses to the events, we can confidently say that if one has any confidence in history whatsoever, then they have little to no reason to doubt, (or be agnostic to), the historicity of the resurrection of Christ as the best explanation and hypothesis to the events surrounding Christ's death, empty tomb, post mortem appearances, and the very origin of the Christian faith.

One might still say: *"well, I don't think miracles are a very good explanation,"* but we must remind them, that type of statement is a philosophical statement, (which they are completely entitled to hold), and that is irrelevant to what we know concerning the historic facts around Christ's crucifixion. As resurrection critic Wolfhart Pannenberg admitted to William Lane Craig:

> *"The facts that an event bursts all analogies to the present cannot be used to dispute its historicity. When, for example, myths, legends, illusions, and the like are dismissed as unhistorical, it is not because they are unusual but because they are analogous to present forms of consciousness to which no historical reality corresponds. When an event is said to have occurred for which no present analogy exists, we cannot automatically dismiss its historicity."* [116]

Much like Pannenberg admitted, we must look at the facts with an unbiased approach, and when we do, skeptic and agnostic alike, must come to the acknowledgement, that all arrows do squarely point and attest to the resurrection of Christ as the best hypothesis. Whether it is true or not is irrelevant to the strict historian, because he or she is simply reporting the facts as they stand; much like it is not the historian's responsibility to rule out Alexander the Great never losing a battle because of its very unlikely hood, (none of the great military leaders throughout history have ever died undefeated in battle and made such an impact as Alexander), nonetheless, if we interpret the facts with an unbiased approach, we must accept the fact that Alexander the Great was undefeated in battle, and moreover, that the resurrection of Jesus by far stands out as the best hypothesis surrounding the three facts mentioned above.[114]

> "No historian can legitimately rule out documentary evidence simply on the ground that it records remarkable events. If the documents are sufficiently reliable, the remarkable events must be accepted even if they cannot be successfully explained by analogy with other events or by an a priori scheme of natural causation."[117]

What is so unique about Christianity and Christ?

At this point, I would like to follow along my good friend Dr. Subodh Pandit's same journey of inquiry that he describes in his search seminar series and book; and thus quickly tie this in with other religious thoughts, and how they match up in contrast. While it is not my place to judge anyone, I will simply let Dr. Pandit lay out the facts and remind each of us that any religion not willing to examine itself will gain a following only through intimidation and fear.[118]:

- Hinduism's highest claims: No single founder, but the highest claim is to be that of a sage
- Islam – Mohammed was called the "Seal of the Prophets."
- Buddhism – Super-enlightened one, who was enlightened in stages while under a fichus tree in India.
- Christianity – Jesus didn't claim to just show truth, but to actually be truth as the Son of God.

Obviously all of these are very prestigious roles, but only one is an "out of this world" claim. **All are recognized as being in the grave except Jesus.**
- Only Buddha witnessed his enlightenment.
- The authors of the Hindu Scriptures wrote what was handed down to them. They did not experience the actual stories themselves.
- Only Mohammad saw the angel Gabriel and interpreted these visions. Islam's claims are solely based on Mohammad's witness.
- Christianity hosts at least 40 authors through vast periods of times, from kings to peasants, all parallel in their complete message being fulfilled by the one prophesied of in Jesus Christ. The Apostles and

Disciples of Christ did not hear about Christ, they actually walked with and lived with Him first hand.

All major religious figures had to find their journey.
- Buddha had to search for enlightenment
- Mohammed was revealed the word; he did not have it, it had to be revealed to him
- Jesus knew from His birth what His purpose and mission was.

All religious texts state that their founders sinned except for Jesus, which was claimed to be without sin.

The beginnings of the great leaders, their ministry, and death differed.
- Hinduism – While there is no single founder; all appear to have been revered individuals with honor and prestige surrounding them.
- Buddhism – Gautama Buddha, was a prince.
- Islam – Mohammed was of the tribe of Quraysh, which was very important because it was the guardian of the Kaba the most sacred site in the peninsula.
- Christianity – Born in a stable among farm animals.
- Christ was the only one born in poverty-stricken circumstances. He is also the only one born a virgin birth, which both the New Testament and the Koran testify to.

The leaders' deaths were quite different as well.
- Hinduism – Their deaths were regarded as significant losses, to be lamented over for generations, as their lives were recounted over and over again.
- Buddhism – Buddha's body was wrapped in one thousand layers of finest Benares cloth and cremated. He was honored and his followers left to continue his work of searching for enlightenment.

- Islam – Umar, one of the prominent leaders, hurried ashen-faced to the mosque to mourn, and express their great respect and admiration. Mohammed had built the framework for the juggernaut of the caliphate that would later sweep the world and bring almost every opposition to its knees.
- Christianity – Was tried as a criminal; tortured, beaten and convicted and died with two robbers. His disciples having abandoned him, he died as a criminal.

When we look at the differences from a number of perspectives, it is quite amazing how the differences cry out. Why I would never purposely mean to disrespect anyone's tradition or religious leaders, we can see that Christ stands out as "not of this world." Both in His life of non-materialism, non-earthly empire, and non-prestigious death; but moreover in His resurrection.

It is quite amazing when you piece together what philosophers call a cumulative case argument, where you put all the facts together to form the best hypothesis, how Christ stands out so predominantly. Even my Muslim friend does not deny this. It is amazing from these poor surroundings, in an area of the world not deemed particularly important, that this Jewish Rabbi, makes sense of all things in and out of the world.
- Christ's resurrection story grew rapidly in the very town where He was crucified.
- His Brothers did not believe in Him fully until His resurrection, when they became so devout in their belief, that they died rather than deny Him.
- All the apostles except for John were martyred for their faith. All they could get from following the Lord, was torture, persecution, and eventual martyrdom, but they followed and never doubted Him again after the resurrection.

- Hindus consider Jesus a God/Avatar, Buddhism hold Him to be a true model to follow and that He did reach the ultimate stage of Nirvana; Islam holds Him to be the Messiah, that He lived a pure and righteous life, was born of a virgin, and that He will come back at the end of the age; Christianity holds him as the invisible expression of God Himself who died for our sins, and will come back at the end of age; atheists/agnostics hold Him as a model moral teacher.

It is indeed amazing that one person in the history of the world stands out like this individual.

*Nothing I have discussed in this chapter is overly controversial, and the majority of even skeptics will agree with my points. (They may differ on my conclusions, but the facts speak for themselves)

Moreover, nothing I have discussed in this chapter is that difficult or complicated to learn; therefore, there is no reason why all Christians do not at least know these basic facts about God, Christ, and the faith they profess; in accordance to 1 Peter 3:15: "Always be ready to give an account to anyone that asks you for the reason of your faith, with humbleness and respect." This is going to be increasingly important in an increasingly secular world and if we want to have any chance of providing all facts to our young adults and future leaders. So as my good friend Dr. Pandit concludes in his book: "May the God of Truth guide you and bring you safely to the harbor of fulfillment and meaning in your life."[118]

The most important thing that we must not let ourselves lose track of is that of having a personal relationship with Christ. This is the true "Theism" of Christianity. The most important part of this journey is to know and accept the person of Christ. While there is nothing I can say to convince you of this, I can testify that the experiential relationship I have with Christ is real, and I encourage you to explore the realness of this for yourselves; for Christ promises if you will just ask Him to make Himself known to you, He will. I pray that you will have this personal acceptance and relationship with Him, and understand what that means for you personally.

So basically, what we are looking at is while secular philosophies and interpretations of science change regularly to say the least, Christ and the Bible have not. What is more surprising in some ways, is the fact that the Bible, while remaining unchanged, has given us an accurate interpretation of the world around us. From the beginnings of space and time in what is now recognized as the "Big Bang," the rapid creation of life, (often referenced in the Cambrian Explosion), the rapid sophistication in which humans came into existence in the approximate regions of Samaria/Mesopotamia, history as we know it, all climaxed with time literally being divided with Jesus Christ coming into the scene (BC/AD); all match the biblical narratives, and moreover, the natural and moral theology all around us. So what do we expect to see in the future? If history repeats, naturalistic theories will come and go, with atheists continuing to claim, "We have it figured out now!" while the Bible and Jesus' words remain the same and timeless; and once again, we will reflect on the truth of Robert Jastrow's assessment which he makes in his book, "God and the Astronomers:"

"It is not a matter of another year, another decade of work, another measurement, or another theory; at this moment it seems as though science will never be able to raise the curtain on the mystery of creation. For the scientist who has lived by his faith in the power of reason, the story ends like a bad dream. He has scaled the mountains of ignorance; he is about to conquer the highest peak; as he pulls himself over the final rock, he is greeted by a band of theologians who have been sitting there for centuries." [67]

Conclusions

In summary, I will use William Lane Craig's 5 premises and add a sixth – we should all be willing to acknowledge these as good reasons to think that the God of Christianity exists:

1. God makes sense of the origin of the universe
2. God makes sense of the fine-tuning of the universe for intelligent life.
3. God makes sense of where/how all information originated
4. God makes sense of objective moral values in the world.
5. God makes sense of the life, death, and resurrection of Jesus Christ.
6. God can be immediately known and experienced

These are only a very small part of the evidences for God's existence of course. Alvin Plantinga, one of America's leading philosophers, has laid out two dozen or so arguments for God's existence. Together these constitute a powerful cumulative case for the existence of God.

Now if we are travelers and not merely balconeers, the conclusion that God exists is but the first step of our journey, albeit a crucial one. The Bible says, "He who would come to God must believe that he exists and that is a rewarder of those who seek him" (Hebrews 11:6). If we have come to believe that he exists, we must now seek him, in the confidence that if we do so with our whole heart, he will reward us with the personal knowledge of himself.[119]

What is Your Wager?

That ties in with Blaise Pascal's famous "wager" know as the Pascal wager. Pascal was a French mathematician and physicist who was the founder of the probability theory, who came to acceptance of Christ in 1654. The Christian religion he claims, teaches two truths: That there is a God whom men are capable of knowing and that there is an element of corruption in men that renders them unworthy of God. Knowledge of God without knowledge of man's wretchedness begets pride, and knowledge of Jesus Christ furnishes man knowledge of both simultaneously. Pascal invites us to look at the world from the Christian point of view and see if these truths are not confirmed.

The human predicament leads up to his famous wager argument, by means of which he hopes to tip the scales in favor of theism. Pascal argues that the prudent man will gamble that God does exist. This is a wager that ALL MEN AND WOMEN MUST MAKE – the game is in progress and a bet must be laid. There is no opting out; you have already joined the game. Which then will you choose – God exists or He doesn't? If one wagers that God exists and he does, one has gained eternal life and infinite happiness. If he does not, one has lost nothing. On the other hand, if one wagers that God does not exist and He does, one has suffered infinite loss. Hence, the only prudent choice is to believe.[120]

Now Pascal does believe that there is a way of getting a look behind the scenes, to speak, to determine rationally how one should bet, namely the proofs of Scripture of miracle and prophecy, (which we discussed earlier). So it is not a blind leap of faith which like many who usually have not read or know the full details of the argument may think; rather it is like looking at a bridge going from one side of a cliff to another. Is it safe to cross? What this chapter will hopefully accomplish is that like Pascal suggests, if we study the bridge, we see a motorcycle cross it safely, then a car and finally a large truck crosses it safely. It still takes "faith" that the bridge will secure us, but it is anything but "blind faith." Therefore, (much as Pascal concluded), it takes more faith to convince yourself the bridge will not support you (atheism) vs. the fact that you have every reason to believe the bridge will support you; and moreover, that to stay on one side equals ultimate destruction while the other side could "possibly" really lead to ultimate safety and reality. Why would you even think about not crossing the bridge?

Bridge that my wife and I crossed in the Highlands of Scotland in 2008; just like the steady bridge of faith that we find in Christ, we had no lack of faith that this bridge was more than secure enough for us to cross it.

So while I feel that just these few facts that I have reviewed should be seriously contemplated by the skeptic, and engrained and researched by the believer so their faith will be strengthened and they will be able to defend why they believe in Christianity; it is really between you and "God" in my opinion. But as Pascal points out, the game is in progress and one must decide what they will do with their lives and with this God. Which road will you take?

Alice came to a fork in the road. "Which road do I take?" she asked.

"Where do you want to go?" responded the Cheshire cat.

"I don't know," Alice answered.

"Then," said the cat, "it doesn't really matter which road you take, does it?"[121]

So I leave you with that. What road will you choose my friends? What is your bet?

> *"It is good to have a reminder of death before us, for it helps us to understand the impermanence of life on this earth, and this understanding may aid us in preparing for our own death. He who is well prepared is he who knows that his is nothing compared with God, who is everything; then he knows that world is real."*
> *Black Elk – Oglala Sioux*

[20] *Lo! I stand at the door, and knock; if any man heareth my voice, and openeth the gate to me [if any man shall hear my voice, and open the gate], I shall enter to him, and sup with him, and he with me.*
(Wycliffe New Testament)
(Rev. 3:20)

CHAPTER III
ඏ෴ඏ

*"I like your Christ. I do not like your Christians.
Your Christians are so unlike your Christ."*
Mahatma Gandhi

<u>What Gandhi a "Christian?"</u>

"**G**andhi a Christian" you are probably exclaiming! Well let's begin by simply looking back to the origin of the word "Christian." Christian was first used more as a derogatory term for those that were so devoted to the person named Jesus of Nazareth "The Christ," that these disciples so attempted to duplicate Christ's actions that they began to be referred to as "Christ-ians." So with that in mind, let's quickly take a few quotes from Mahatma Gandhi and compare those as well as Gandhi's actions with that of Christ's, and I will leave it to you to decide, and more importantly, to hopefully learn from. We all know the vast influences that Gandhi's legacy has had on the entire world. He faced stiff resistance, but insisted on a non-violence approach to any and all forms of protest. "I have nothing new to teach the world. Truth and nonviolence are as old as the hills" he said. So if we look at his compassion for the down trodden, his emphasis on surrendering to "God," his self-sacrifice, and his life search for "absolute truth," (he even had a picture of Christ in his room and knew the New Testament very well), so I think we can all come to an agreement, in this light that a case could be argued that in some ways he did follow Christ; but what we all really are most concerned with, is if he is a "self-proclaimed" Christian and what church did he belong? Am I wrong?

> *"Is Mr. Gandhi a Christian?" a visitor asked Millie. Ms. Millie asked for further clarification whether she meant one converted to Christianity or one who believed in the teachings of Christ. The visitor emphatically told she meant former. She was talking about him with some friends and they were wondering because Gandhi knew Christian scriptures so well, and was fond of quoting words of Christ frequently and hence her friends thought he must be a Christian.*[174]

When the missionary E. Stanley Jones met with Gandhi he asked him, "Mr. Gandhi, though you quote the words of Christ often, why is that you appear to so adamantly reject becoming his follower?"

Gandhi replied, **"Oh, I don't reject Christ. I love Christ. It's just that so many of you Christians are so unlike Christ. If Christians would really live according to the teachings of Christ, as found in the Bible, all of India would be Christian today,"** he added.[175]

Similar to today, Gandhi referenced repeatedly in his writings and conversations that if it wasn't for such fake and un-authentic Christians, him and all of India would be Christian. So to give a quick comparison, General Dyer of the British army ordered British troops to fire on men, women and children killing just over 1,500 during Gandhi's life. General Dyer was a self-proclaimed Christian, so while

I am ardently against pluralism/relativism, I will leave it up to you to decide who is the true "Christ-ian" in these two examples. India has over 1.1 billion people today, and will soon surpass China as the most populated country in the world. Can you imagine what India might look like today if Gandhi's words were right: "<u>If Christians would really live according to the teachings of Christ, as found in the Bible, all of India would be Christian today</u>." Gandhi experienced Christianity through Western meditation: words, actions, and writings. But when he read the New Testament he felt there was a huge gap between the message he found there and the behavior of Christians in the West. In fact, he struggled with its Westernization because the country where Christianity originated, Palestine, is part of Asia.[176]

It seems that more and more cultures and people groups are agreeing with Gandhi's assessment of Christ vs. Christian. But is this a fair statement? Is it ironic, however, that Christianity's history is laced with division and self-destruction, politics, hate and judgment, and internal strife; while Christ represented the exact opposite way for us to follow.

While this will seem very harsh, I feel inclined to state what we all know to be mostly true; that the vast majority of "Christians" of the past and of the present are not truly followers of Christ, and I think we all know this, (including myself). This, however, should not be a point of utter despair, but a point of reform and change; change putting Christ and His teachings in our very being, and the center of every aspect of our lives. We must acknowledge, however, that proclaiming the title "Christian," is not enough. We must ask ourselves "is Christ truly in us?"

I would like to apologize on behalf of everyone ever hurt by those claiming to follow Christ in words, but not in actuality, including myself. I apologize also for those who felt a genuine connection with Christ, but had it stifled because it did not follow along an institutionalized or organized pathway. For example, many of us now acknowledge that many of the Native Americans were more "Christian" than the "Christians" who came to North America to westernize them. While I am less than 10% Native American myself, I spent two years with the Cherokees and worshipped with them on a regular basis from 2000-01. What I found was a true connection between Native American tradition and Christ. We must remember that to be Christian is NOT to be Western. Scripture tells us that we first have general revelation of nature and the world around us to connect with God; secondly we have conscious revelation through morals, etc, and finally we have divine revelation. So just as many of the Old Testament prophets were saved thru Christ, though Jesus of Nazareth was yet to be born; Native Americans too were in many instances more "Christian" than those who sought to convert them.

In my opinion we can still learn much from these Native Americans. They could have better taught us in many ways how to be in a better relationship with the Great "Holy" Spirit, than many of us are at present during our ONE hour per weekend church service.

> *"Christianity is a religion, a system of faith and worship. It appeared to the Indian that the followers of Christianity donned their religion only when the need arose. They spoke of beautiful philosophy and spent entire days in prayer, but, the following day, all these things lost their meaning, whereas Indian faith and worship was a daily lifestyle. Christianity seemed to be a religion of supplication, whereas, Indian thoughts were praises of thanksgiving. Christianity did not provide a place for the Indians to conduct their expressions of thanksgiving, which*

were customarily carried out in ceremonials of song, dance, and feast."[123]

I hope we can see from the above quote, that this type of thought vs. a forced westernization of a people group was truly more "Christ-like" than by many self-proclaimed "Christians." I still experience this in many churches where there are certain "rules" to not show emotion during singing praises to God; whereas when I worshipped at a non-denominational Cherokee church with a Cherokee pastor in 2000-01, we had the restrictions lifted, to keep the focus on Christ. This is another reason why many Native Americans today reject Christianity, because of the Christians forcing of a westernized version of it. The more I study Christ and Theology, the more I see that our roles are to promote a Christ following and not worry so much about "church following," and leave the rest to the Holy Spirit, (even though I do think church membership is an essential part of fellowship). Therefore, if I have developed a friendship with a Native American who practices his or her own Native American beliefs as outlined above, my role is to show how Christ is this accumulation of their Great Spirit, so that if they walk away saying they're a Native American follower of Christ, or a Messianic Indian, I should be quite pleased in leaving the remaining details to the Lord. (Please don't misunderstand where I am leading; my present graduate work in Theology is from a traditional oriented school, and I have learned a great deal of drawing closer to Christ through my present church and a traditional study, but we must remember the simple fact that Christ did not preach for institution, instead he instructed us to remove the shackles and worship God in "entirety" with all of our being. (Whether we like it or not, Christ was a type of revolutionary)).

When we are talking about Christians throughout the world; Messianic Native Americans, Messianic Jews,

Messianic Muslims, Messianic Buddhists, Messianic Hindus, Messianic Christians – this means that while my wish is for all to accept the Bible from cover to cover, we must first start with people where they are, and the starting point, ending point, and whole of the Bible is Christ. Christian is just a title, just like Messianic Jew is; what matters is the relationship that particular individual has with Christ. (If we are to be honest, how many of us in the United States would have to be classified as "Messianic secularists?" We follow Christ, but we also follow secularism in our jobs, hobbies and education as well.) The December 2009 issue of Christianity Today had a very interesting article on this exact same point:

> *"Ever since the Wesleyan revival and the Great Awakening of the 18th century, evangelicals have insisted that what matters most to God is not one's identity as 'being Christian,' but rather whether one has a life-transforming relationship with Jesus Christ. –With Messianic Jews, the evangelical community mostly accepts that the label 'Christian' is not essential. –Jesus said that 'whoever comes to me I will never drive away.' As Nabil and Ibrahim (persecuted Muslims who follow Christ but still consider themselves Muslim), understand their position in the universal body of Christ, they must listen to counsel from others around the world. But if we understand our position in that same body, then we must respect their fundamental human right to sort out –under the authority of Scripture—how they express their identity as followers of Christ. It is they whose lives are quite literally on the line. If they can respect each other after suffering prison for Jesus, then surely we can treat them both with respect."* [124]

You are free to interpret this how you would and to draw your own conclusions, but I do hope that no one will doubt that our first role is to love all (as Christ would), be in relationship with them, and tell them about Christ and the heart of the Gospel. Then if someone as mentioned in the above question decides they will stick to their mosque and not your church, but that they have decided to follow Christ, then we should say "Praise the Lord!" If they are following Christ, then we need to have faith in Christ that He will take care of the rest. What I am getting at is that while there are thousands of denominations and it is highly unlikely that every church that professes they are the "chosen church," are really the only chosen church; we need to concentrate on our similarities (namely on JESUS CHRIST) and less on the differences that split churches such as musical instruments or who has the biggest ego, etc.

Christ Himself gave a similar example for us and our churches to follow on the subject:

> *"'Teacher,' said John, 'we saw a man driving out demons in your name and we told him to stop, because he was not one of us.' 'Do not stop him,' Jesus said. 'No one who does a miracle in my name can in the next moment say anything bad about me, for whoever is not against us is for us. I tell you the truth, anyone who gives you a cup of water in my name because you belong to Christ will certainly not lose his reward.'" (Mark 9:38-41)*

So I would assume that Christ would tell a Messianic Muslim, that He loves them and died for their sins too. Correct? I disagree with Islamic theology 100% as mentioned earlier, but we must be realistic (as I have discovered), that it is unlikely a Muslim (or any other believer) will drop all of their previous beliefs all at once. Therefore we must start with Christ as Paul did:

> *"I planted the seed, Apollos watered it, but God made it grow. So neither he who plants and the man who waters is anything, but only God, who makes things grow." (1 Corinthians 3:6-7)*

More or less denominations in the future? (you decide)

"One hundred religious persons knit into a unity by careful organization do not constitute a church any more than eleven dead men make a football team. The first requisite is life, always."
- A.W. Tozer

This is the question that throughout history up until the present day (and will play a large factor in our future), has plagued the church in many ways. I have witnessed churches splitting on doctrines that may well justify a parting of ways over a pollution of Scriptures (example – Episcopal Church); but I have also seen churches parting because of one not using the King James version of the Bible, disagreement over the use of musical instruments, PowerPoint, and so on. With this in mind, let us remember with humility that if we follow Christ, we are of the same body (church); Paul sums up this diversity of believers and our role very well in 1 Corinthians:

1 Corinthians 12

>[12] The body is a unit, though it is made up of many parts; and though all its parts are many, they form one body. So it is with Christ. [13] For we were all baptized by[a] one Spirit into one body—whether Jews or Greeks, slave or free—and we were all given the one Spirit to drink.
>
>[14] Now the body is not made up of one part but of many. [15] If the foot should say, "Because I am not a hand, I do not belong to the body," it would not for that reason cease to be part of the body. [16] And

if the ear should say, "Because I am not an eye, I do not belong to the body," it would not for that reason cease to be part of the body. [17]If the whole body were an eye, where would the sense of hearing be? If the whole body were an ear, where would the sense of smell be? [18]But in fact God has arranged the parts in the body, every one of them, just as he wanted them to be. [19]If they were all one part, where would the body be? [20]As it is, there are many parts, but one body.

[21]The eye cannot say to the hand, "I don't need you!" And the head cannot say to the feet, "I don't need you!" [22]On the contrary, those parts of the body that seem to be weaker are indispensable, [23]and the parts that we think are less honorable we treat with special honor. And the parts that are unpresentable are treated with special modesty, [24]while our presentable parts need no special treatment. But God has combined the members of the body and has given greater honor to the parts that lacked it, [25]so that there should be no division in the body, but that its parts should have equal concern for each other. [26]If one part suffers, every part suffers with it; if one part is honored, every part rejoices with it.

One thing that Ravi Zacharias fears about evangelical Christianity that conveys my point, is a fear that some feel they have to completely destroy every aspect of one's present religion or mode of thought, and replace it with their own. Paul realizes and speaks to his various audiences, first on praising them for their yearning of God and for what they have right, and then filling in. Paul likewise commended the people for their search; similar to Jesus commending the woman at the well for the knowledge she did have. In the framework of many people's beliefs, they are not wholly wrong; their system may be wrong, but rarely are they wholly wrong. When we are trying to reach someone we need to approach them in this same manner. Likewise, if a Muslim praises Jesus as a great prophet and the Messiah, a Hindu says that Jesus is a God of many Gods, or a Buddhist states that Jesus reached the karma of ultimate enlightenment – we have a great foothold on which to build, and they are obviously not wholly wrong, so there is no advantage for them or for the gospel, to attempt to tear down the entire foundation. Ravi (nor myself) are endorsing a joining of religion thought by any means, but simply encouraging the approach of Jesus and Paul – to reach the people where they are and build on their foundation. What better way to build on that foundation, than on that of Jesus Christ? Hence – a "Messianic Muslim, Jew, etc," is not a bad starting point at all. Thus, as Paul pointed out to the people of Athens in Acts 17, (who were even worshiping idols and a multitude of gods), I would suggest as a good starting point for us to follow. He complimented them, and then reached them where they were, building on the level of religion that they were currently at: "Men of Athens! I see that in every way you are very religious. For as I walked around and looked carefully at your objects of worship, I even found an altar with this inscription: TO AN UNKNOWN GOD. Now what you worship as something unknown I am going to proclaim to you." Let us too, build on that foundation with the Gospel of Christ, (and less on our western interpretations).

We must remember that Christ Himself said one such as a prostitute would inherit the Kingdom of Heaven ahead of the overly religious Pharisee; is it not ironic how we could pride ourselves in being superior to a poor Eastern Christian, a Messianic Muslim, or Messianic Jew, who each profess faith in Christ; even though our own Bible warns us against this type of condemnation. Do we remember who acknowledged the following:

- The serene faith of a pagan centurion (Matt 8:10)
- The persistent faith of a Canaanite woman (Matt 15:28)
- The desperate faith of the thief on the cross (Luke 23:43)
- Samarian woman at the well. (Luke)
- Chose common fisherman instead of the religious leaders

(In case you didn't guess it, it was Jesus who acknowledged each of these)

We are called to be the "salt of the world," but this does not mean through the extremism view of political correctness, compromise, and conformity of our very beliefs. We have explored many of these examples already, but I can not harp on them enough for the future of Christianity. We must always place Christ at the center of our very being; and moreover if we are His Disciples, we must engage the world as "salt and light."

While we need to be prepared to openly challenge the educational systems that have admittedly falsified information when certain facts point to God or Christianity, and challenge the same system that will not allow a high school valedictorian or university graduate to publicly acknowledge Christ; we must remember as the book "Re-Jesus" states: "It seems to us that a constant, and continual, return to Jesus is absolutely essential for any movement that wishes to call itself by His name."[125]

By keeping Christ in the center, would we not have avoided the subject mentioned in the first chapter, concerning internal strife, which resulted then in excommunication of different Christian factions from one another? While we do not have excommunication in modern times, we do have division and strife on such topics as musical instruments, which is the correct Bible version, etc., that I have seen some Christians abandon their Christian Brothers and Sisters on such minute matters as these. I must ask again, does anyone really believe that this is what Christ would want? To be divided over one church preferring only a piano, while another prefers drums, while another prefers no music; I think not. What I must plea for is a form of the "scary" word known as Ecumenism. What I mean by Ecumenism is those core doctrines that we all agree on. What C.S. Lewis called "Mere Christianity," focusing on the basic foundations of our Christian faith.

Unity or Division – What would Christ preach?

Ëcumenism – (not "pluralism") – mainly refers to initiatives aimed at greater religious unity or cooperation. Most commonly, however, *ecumenism* is used in a narrower meaning; referring to a greater cooperation among different religious denominations of a single one of these faiths.

The word is derived from Greek οκουμένη (oikoumene), which means "the inhabited world", and was historically used with specific reference to the Roman Empire. Today, the word is used predominantly by and with reference to Christian denominations and Christian Churches separated by doctrine, history, and practice.

Let me ask you a simple question – Eastern Orthodox, Roman Catholic, Protestant – do we all believe in God? What about Christ? That we should follow the teachings of Christ? That we should reach out to a dying world? That we should encourage fellowship and love, like that displayed by Christ? That we need to realize that grace is God's free gift that Christ paid on the cross for us? That Christ's blood has paid for our sins, and that we should accept this free gift of salvation?

Okay – stop right there! If we truly are Christ followers, then we are ALL UNITED on these basic points! It is time that we quit saying how: *"we're the best! We're the chosen! Only with our "church" membership can you get into heaven!"* – And it's time to start saying how: a Billy Graham compared to the drunk on the street are exactly the same; we are all sinners in need of a savior, and that Savior is Jesus the Christ.

I beg of you, please let us set aside our differences in doctrine, and focus on the heart of the matter – worshiping God with all of our heart, making sure Christ is really and truly in me and that other people see it, and reaching out to a world just like Christ did, telling them of the good news; not by being hypocritical or condescending and telling them they must "join our church," but by being their friends, just like Jesus and making sure they know who Christ is. *The rest will come by God's Holy Spirit.

While I will readily admit that I disagree with a very large percentage of Roman Catholicism and Eastern Orthodoxy, I must ask if all of us can please come together on the centeredness of Christ? Are we "One Body in Christ," or a divided body? What do you think Christ would say? I think we all know… We are living in a post Christian west, so it is imperative that we put our differences aside and put Christ in the center. – We of course have differences in music, clothing, baptism criteria, which translation of the Bible to use, etc., etc., etc. but this needs to be kept behind closed doors, while the heart of the Gospels, the Bible, and Jesus the name above all others, needs to be kept anywhere but behind closed doors. Will you please at least consider setting your differences to the side and placing our churches and our personal **focus on the unity of Christ**?

Monument in the Highlands of Scotland – "United We Conquer" – words to live by for Christ followers from an Ecumenical standpoint. We must be united with Christ if we are to have any hope into the 22nd century.

Pastor Francis Chan, (senior pastor of Cornerstone Church in Simi Valley, California and founder of Eternity Bible College), eloquently sums this up in his blessed book on the neglecting of the Holy Spirit in our lives, Forgotten God pg 47: *"Don't let your views be determined by a particular denomination or by what you've always been told. Within the context of relationship with other believers, seek out what God has said about His Spirit. Open up your mind and your life to the leading of the Spirit, regardless of what others may think or assume about you. Fear has a way of channeling our thought process. Fear of stepping outside of a certain theological framework causes us to be biased in our interpretations. We work diligently to 'prove' that our presuppositions were correct rather than simply and honestly pursing the truth."*[155] (Chan's long-term plan involves building the church without having a building, (which will be more crucial for Christianity's true disciple making and growth into the 22nd century)).

Furthering these points is Chinese believer and Brother Peter Xu Yongze who has spent many years in prison for preaching the Gospel in Communist China. He sums up my thoughts on the importance of what many churches continue to label a negative word: "Ecumenism," by referring to the great western missionary Hudson Taylor, (who adopted Chinese customs, and lived with them to better reach the Chinese with the truth of the Gospel, (at the protest of many American churches). Brother Yongze commented:

> *"Another legacy from Taylor to today's Chinese church and to the Back to Jerusalem movement was his refusal to construct organizational walls that slow down the advance of the gospel and cause division among believers. He was able to cross denominational barriers and lead the body of Christ*

> *to work together under the banner of Christ, so that the church could fulfill God's vision."* [153]

His next comments struck me as quite interesting and reflective on today:

> *"In the 1970s and early 1980s, we were all one church. There were no significant theological differences between us and no blockages to our fellowship. We had all been deeply touched by God's Spirit. We went to prison together, were beaten together, and bled together. We also preached the Gospel together. But by the early 1990s many leaders who were once close brothers in the work of the gospel had become critical of each other. We had always stated that we in China would never form denominations like those in the Western church, yet we had all gone different ways and built walls between us."* [153]

After many churches began attempting to reconcile their differences, and get the focus back on Christ, his comments are exactly what I would like for our own denominations to think about:

> *"I stood up in the first unity meeting and said: 'We don't want to follow our own pet doctrines any more. We want to learn from one another's strengths and change in whatever way the Lord wants us to change in order to make us stronger and closer to Jesus.' Although not all differences were ironed out, the leaders came to appreciate their fellow leaders and saw they had far more in common than they had reasons to separate. They also found that their theological differences centered on things that weren't essential to the faith."* [153]

Wow... He is summarizing my point exactly! This is what I mean by "Ecumenism."

Brother Xu Yongze then goes on to say: "When believers are united around a common goal, we can head there together, putting aside our petty differences. When we lose sight of our common vision, we stop looking forward and begin to look at each other. Soon we see each other's weaknesses and faults, and instead of fighting for the kingdom of God we start fighting each other." [153]

Similarly, theologian Wayne Grudem sums it up this way in his magnificent work "Systematic Theology:"

"It is ironic and tragic that denominational leaders will so often give much of their lives to defending precisely the minor doctrinal points that make their denominations different from others. Is such effort really motivated by a desire to bring unity of understanding to the church, or might it stem in some measure from human pride, a desire to retain power over others, and an attempt at self-justification, which is displeasing to God and ultimately unedifying to the church?" [154]

Again – food for thought. I just ask that we all simply ask if forming denominational walls of separation, or keeping the focus completely on Christ is what Christ Himself would instruct us to do. Would Christ take the "church" to the world, or wait for the world to come to the "church building?" I believe we all know the answer to this simple question; whether we follow it or not, is another matter.

"How good and pleasant it is when brothers live together in unity!" Psalms 133:1

We will always have some differences of course, but we should never allow these to be visible. When someone sees a Christian, they should know they have just met a "Christian," because of the light of Christ that radiates from us. "Let your light shine before all men," Christ tells us. Moreover, we can all agree on Christ and the Gospel message, so I pray that the Body of Christ will focus more on this main point of "Mere Christianity" in reaching out to the world; I believe this is beyond critical for the survival of the church into the 22nd century. Ironically the poor Chinese Church of the "Back to Jerusalem" movement mentioned above can teach us a lot in their simplicity (pg 27):

"This is an inter-denominational but not an anti-denominational group of workers accepting the whole Bible as God's revelation. Its aim is to join members of the Lord's body in fellowship to consecrate strength and will on the preaching of the gospel in order to be ready for the Lord's return." [138]

(I see nothing wrong with this type of "ecumenical" statement. Do you?)

My friend Steven Khoury, which I first met at a Voice of Martyrs meeting in Bartlesville Oklahoma in 2008, is a Palestinian pastor in Bethlehem. While his father and him suffer constant persecution from both fellow Palestinian Muslims and Jews alike for operating a Christian church in Israel, I want to focus on a wonderful book he wrote called DIPLOMATIC CHRISTANITY, where he pleas with the United States to forget about trying to be politically correct Christians, because the two do not go together. He goes on to mention one of his favorite stories in the Bible on Shadrack, Meashack, and Abednego in the book of Daniel; how by not being politically correct, or worrying about church protocol, etc., they refused to follow King Nebuchadnezzar's orders to bow down to his statue. He then makes the following comments on his time in the United States:

> *"I'll never forget some of the first churches I visited while studying in the U.S. I did not know much about freedom of worship, nor did I know much about how outspoken one can be about his or her faith. I thought church in the States would be the same for me as it always was. I'd either be coming home beaten and bruised, get my Bible ripped from my hands, or be forced to worship in secret for fear of nosy neighbors.*
>
> *What I found was a whole different set of problems. I remember going to one church where they were remodeling and were squabbling about the color of the curtains and carpeting. At another church I found Christians arguing over who could park where in the church parking lot. Some of the worst experiences I've ever had were in churches that verbally attacked each other over issues like the Holy Spirit, who had the better choir, which praise instruments should be allowed, or whether PowerPoint materials were of the devil. I've also heard Christians argue over specific wording one should use for telling someone about Jesus and which version of the Bible should be allowed in churches. I've actually heard a Christian say that any other version different from his version of the Bible was Satanic..."* [126]

Palestinian Pastor Steven Khoury
(http://www.holylandmissions.org)

Does Pastor Khoury's comments sound familiar? I can attest first hand that in multiple denominations that I have visited, they have unknowingly kept the focus on doctrine more than Jesus Christ and His teachings. *(I should reiterate that I am not in anyway meaning to sound as if I am "church-bashing," I have learned much at the church and they are full of loving Christians; I am just pleading for a quickly declining church to wake up before it is too late, and focus on the simplicity of Christ, while rejuvenating our churches with an urgency of reaching those around us with the genuine/relational Gospel message.)*

While all of these mentioned above are different denominations, I must point out that these churches are full of good loving Christians as well, that would give you the shirt off their back; unfortunately they are predominantly too focused on what you should eat, how you should dress, what type of music should be used, what version of the Bible is in use, how much money you give the church, our church is the "chosen" church, etc., etc., etc. Now the sad thing is, if I asked 90% of these same persons if Christ stood for any of these points, they would say "no." So my emphatic question is "then why are we so focused on them!?!?" I am sorry if this seems to be blunt, to the point, and on the verge of being a little rude, but I must point out the importance of not only following Christ and letting others be drawn to Him because they see Him in us (Christians), but the fact is, that with the average church age being around (or over) 55 years, and the average life expectancy being around 75; it does not take a genius to figure out that in the next 20 years we are going to see a lot of churches closed down. While we bicker about whether some young kid should be able to play a guitar at church, the bigger picture is that the church can't stay opened without members. Does this mean give up our morals? Of course not. What it means is simply follow Christ. If we do this, not only will be truly be "Christians," but our churches will continue to grow.

"The west is becoming hostile towards Christ. This is the peculiar situation of our time, and it is genuine decay. The task of the church is without parallel. The corpus christianum is broken asunder. The courpus Christi confronts a hostile world. The world has known Christ and has turned its back on Him, and it is to this world that the Church must now prove that Christ is the living Lord. The more central the message of the church, the greater now will be its effectiveness."
ETHICS by Dietrich Bonhoeffer pages 109

Please take note of Bonhoeffer's last sentence in this quote – *"The more central the message of the church, the greater now will be its effectiveness."*

I have heard too many young adults say they hear too much "Church-ianity" being preached instead of "Christianity," and I would be lying if I said they were wrong. While I am no subject matter expert by any means, I do have common sense and I can see the "writing on the wall," as well as interpret the fundamentals of Jesus' teachings. Let me use a quick example from my hometown. My home church was an average size church in 1984 and today it is about the same size. A family of 5 started renting out our church when it was not in use the same year. They focused exclusively on Christ and Scripture, while deciding to consider themselves "inter-denominational," and today they have approximately 10,000 members, and have kept the young people focused on Christ, busy with local and overseas mission trips. Of course many will say "they're not picking sides! So they don't stand for anything!" That is my point! They keep their focus on Christ and teach through the Bible. (also of the 10,000 members, probably 40% are 35 years old or younger, (food for thought)). Now am I saying we should "drop denominations?" Of course not, but we must keep the focus on Christ and strive for greater cooperation with other churches and organizations for our future. Another quick example would be a family member's Baptist church which holds close to 500 in their primary worship sanctuary, and at least another 150 in their Sunday evening study room, and (although it is within approximately 1 mile of a large University campus), it now has an average attendance of 30. Realistically how much longer can this church stay opened? If they are within a mile of a university that has approximately 20,000 students from all over the world, then why do they not have the church occupied with something 7 days a week? I simply ask you to reflect on if your church is following doctrines or Christ more, and then make the necessary changes to make certain it is Christ, and stays Him.

In "They Like Jesus, but not the Church," Dan Kimball discusses his conversation with a young adult named Maya, who made this comment about Christianity:

"I actually would want to be told if I am doing something that God wouldn't like me to. I want to become a better person and be more like Jesus. But that isn't how it feels coming from Christians and the church. It feels more like they are trying to shame you and control you into their way of thinking and personal opinions about what is right and wrong, rather than it being about becoming more like Jesus and a more loving human being." [127]

Does this sound like you or your church? Whether it does or doesn't, I know it doesn't sound like Jesus, and I pray that this young adult was given proper Christ centered advice after she made this comment. As Dan goes on to point out, we can teach our churches and one another to be salt and light to the world, by changing the image of the church from negative and judgmental (opposite of Jesus) to the church is a positive agent of change, loving others (including outcasts) as Jesus would. It all starts with one person.

We do need to first admit that there is a problem in the churches of the west. Second, is get our focus back on Christ, (this does not simply mean nodding your head and continue doing the same thing). It means first that we must focus on being "Christ Followers," over "denomination followers." I recently saw a group of people at a busy intersection simply handing out bottles of water with their church website on them. The bottles read something to the effect "we all are thirsty." That's it. Because they followed Christ's example, of simply letting their light shine without immediately promoting their specific church or doctrine, they are finding themselves very successful and their membership growing. (They also made it a point to go to a Laundromat and give quarters to everyone there to pay for their laundry). Again – if we ask ourselves, is this the type of example that Christ would like to see us doing, I think we can begin to figure out things quite quickly. I recently was working to organize a gifted apologist Dr. Pandit from India to appear at 2 local universities and end at our church; the church board was so concerned about their denomination being promoted, etc, that I finally canceled the whole thing, and will continue to do as I have in the past, and that is support the cost myself (with the Lord's guidance), and get as many friends and family as I can to help me get the urgent message of the particular guest speaker that I am working with to the people outside of church that desperately need and want to hear.

We must change this mindset, and get it back on the basics of Jesus and the Gospel. I am guilty of this as well, but Jesus Himself went to the ghettos and marginalized areas to show love and tell about the "truth," but if we say today,

"C'mon church members! Let's go down to the local bar strip, ghetto, etc, (not to judge), but to show love and compassion, and that these peoples have a place and friend in us and our church," most will look at us as if we're crazy. I had one church member tell me once, *"uh... we don't do it that way here..."* (After I said why not go the university campuses or even some of the adults visit the local bar strip.) I have simply heard too many young Christians following the path of "I am leaving the church to be a Disciple of Jesus," and I myself have been close to the same point before, but it doesn't have to be that way.

Have you ever heard a young adult say something like *"I don't believe in organized religion."*

If you have, it could be (as I mentioned above), summed up in this comment by 24 year old molecular biologist Alicia:

"Why do I need church? It isn't necessary. I have a relationship with God, and I pray a lot. But I don't see the point of having to add on all these organized rules like the church leaders think you should do. It feels like they take something beautiful and natural and make it into this complex nonorganic structure where you now have to jump through hoops and do everything in the way the organized church tell you. It seems to lose all its innocence when it becomes so structured and controlled."[128]

More and more we are seeing that if we go back to the basics of early Christianity, and form smaller fellowship gatherings, house churches, cell groups, or whatever you would like to call it; we see the Gospel coming alive and spreading just as it did in the Book of Acts. I have personally received more theological study, fellowship, outreach, and felt the Holy Spirit in these smaller groups, and seen them grow rapidly while providing more weekly guidance than just 1 church service per week can do. We need to make sure that we remember that the "church" is the body of Christ, and not just a church building. We must not place a 1-2 hour church service as our foundation, or we will be setting our foundation on sandy soil. Again, I feel we can learn a great deal from our persecuted brethren (such as China) who have to meet and have fellowship in homes or wherever they can. Moreover, we should use our blessings to take "church" to the world, and quit waiting for the word to come to our "church building."

A single Church service on the weekend can not be our "foundation;" it is just a tip of the top of the pyramid. The effects of it being our foundation can be seen with church populations and effectiveness declining sharply.

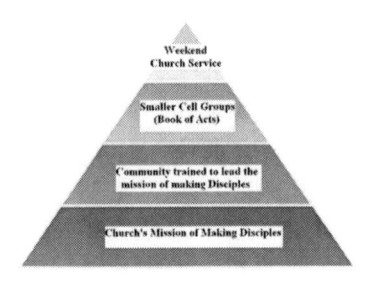

```
        Weekend
      Church Service
    Smaller Cell Groups
      (Book of Acts)
  Community trained to lead the
   mission of making Disciples
  Church's Mission of Making Disciples
```

"Don't emulate the scribes and Pharisees, for they have set themselves up above the masses, and love the honor and praise of men, including being called 'Rabbi.'" [He is cautioning us not to look for exalted titles and ranks, for we are all just brothers in a great, big family—-God's family. And there are no ranks and titles in a family! So he says:] ". . . you are not to be called Rabbi: for you have one teacher, and you are all brethren. And call no man your father on earth, for you have one Father, who is in heaven. Neither be called masters, for you have one master, the Christ"
(Matt. 23:8-10).

I often wonder why we don't think back to the first-century church scene: Who conducted the early communion services? From what seminary did they graduate? What denominational Sanction did they have? Was it not the rough hands of unlettered fishermen who broke the bread of those early days? Or perhaps it was the cleansed heart of a converted publican that expressed thanks for "the blood of the new covenant shed for the remission of sins."

Non-Seminary/Persecuted Chinese Faith Lesson 101

I have mentioned the Chinese and persecuted church in several references earlier in this book, but I think when we study the Book of Acts and see how small cell groups can work in spreading the heart of the Gospel in both word and deed; I can not help but think about applying the following to our outreach (as well as our normal church service), mentality:

"In his book China: The Church's Long March, Adeney joyfully documented the strengths, the Chinese house churches had developed during their years of hardship. The following are some of the most important of these strengths:

1. The house churches are indigenous – The dynamics flow from their freedom from institutional and traditional bondage.
2. The house churches are rooted in family units – The believing community is built up of little clusters of Christian families.
3. The house churches are stripped of nonessentials.
4. The house churches emphasize the lordship of Christ – Because Jesus is the head of his body, the church must place obedience to him above every other loyalty; it cannot accept control by any outside organization.
5. The house churches have confidence in the sovereignty of God.
6. The house churches love the word of God – They appreciate the value of the Scriptures and have sacrificed in order to obtain copies of the Bible. Their knowledge of the Lord has deepened as they have memorized and copied the word of God.

7. The house churches are praying churches – With no human support and surrounded by those seeking to destroy them, Christians were cast on God and in simple faith expected God to hear their cry.
8. The house churches are caring and sharing churches – A house church is a caring community in which Christians show love for one another and for their fellow countrymen. Such love creates a tremendous force for spontaneous evangelism.
9. The house churches depend on lay leadership – Because so many Chinese pastors were put into prison or labor camps, the house churches have had to depend on lay leaders. The leadership consists of people from various walks of life who spend much time going from church to church teaching and building up the faith of others.
10. The house churches have been purified by suffering – The church in China has learned firsthand that suffering is part of God's purpose in building his church. Suffering in the church has worked to purify it. Nominal Christianity could not have survived the tests of the Cultural Revolution. Because those who joined the church were aware that t was likely to mean suffering, their motivation was a genuine desire to know Jesus Christ.
11. The house churches are zealous in evangelism – No public preaching is allowed. People came to know Christ through the humble service of believers and through intimate contact between friends and family members. The main method of witness in China today is the personal lifestyle and behavior of Christians, accompanied by their proclamation of the gospel, often at great personal risk."[144]

I pray that we are beginning to see the value of following the outline given above, that is also outlined throughout the Bible; to build up groups of house churches and cell groups of personal fellowship that study the Word of God in humbleness and prayerfulness, as they seek to grow in understanding and discipleship to Christ. Whether this grows in meeting at coffee-houses, restaurants, bars, clubs, parks, etc, etc, the important piece is growing in Christ in all forms. (Would Christ not agree?)

This is a small and simple church in a village in Nicaragua that I had the privilege of visiting during a medical mission trip in September of 2009. It has neither running water, (nor a door), but is there anything less Christ-ian with this building than a huge mega church? Perhaps we should focus less on debt-ridden large churches that are open only a few hours per week, and instead invest that money into 100 of these smaller type churches (except

maybe with a door), all around the community? (While this may be an extreme example, I think you can start to see my point in the Church being a lot more than some building.) The Body of Christ aka church – needs to go to the community of the world and not wait for the world to come to a church building; it is crucial for "church" to not just be thought of as a weekend 1-2 hour service in a building called "church."

While I have contended that I am in no way against organized religion at all, could we not also learn from this simplicity of focusing on Christ in smaller group settings? I had an experience, (before withdrawing myself from all church offices in June 2009), in being nominated to be an assistant leader to outreach for our community. I soon realized that we all had been disconnected to at least a degree from Christ. We were getting ideas together, and as mentioned previously, I suggested that maybe we consider going to the local bar strip (that many college students as well as all ages visit, (and that I too visited on several occasions as a college student)); not to judge or act better than anyone, but to simply greet them and maybe hand them a Gospel of John booklet or something. I explained that we of course have no reason to judge anyone, but if we simply said something like: "Hey buddy! Did you get one of these? (Handing them a Gospel of John, inviting them to some event or something) Have a great night!" This would hopefully show non-judgmental Christians while not watering down the message, but even wishing them "a good night." Then if someone wanted to talk, etc, they could. I was looked at like I was nuts for suggesting such an idea. Similarly I was asked to be a Deacon again for 2009, but then was told by the Senior Deacon: *"I see you wearing a tie sometimes, and that is good. I sometimes see that one guy that has been nominated as being a Deacon, and he never wears a tie and sometimes even wears his work jacket!? That won't happen with me as Senior Deacon. We can get him a tie and a nice suit of clothes if he needs."* Wow...

My simple question is, "Is this really what Jesus would do or say? Would he really be that worried about what type of tie a person was wearing?" Maybe He would, and I am badly mistaken. But I don't get that from Scripture. (Needless to say, I had to decline being a Deacon for 2009...) Reading this passage by Dan Kimball reminded me of that exact situation:

> *"Jesus spent time with those who weren't religious. He talked with them, he listened to them, he cared for them, he cried for them. He died for them. I wonder if many of us have been so busy inside our churches that we haven't really stopped to observe and listen to those outside our churches as Jesus probably would have. It's so easy to be comfortably numb in our subculture and forget about the horrific sadness of those who have yet to experience the saving grace of Jesus. It's easy to do this when we aren't in personal friendships with them. But I can tell you that when you are in authentic friendships and relationships with those outside the church, you can't forget them. You want them to know Jesus because of what Jesus means to you and how he changed you. We need to have our hearts constantly broken for people, like Jesus' heart was broken. We need to look around us and see people through his compassionate eyes (Matt. 9:36). The question I have for you is, Are you in the prison of the Christian bubble? Have you become comfortably numb? Perhaps you haven't realized it, but you are there. Are you going to surrender to it, or are*

you planning your escape? People like Jesus are waiting on the outside to meet you."[129]

Kimball goes on to gather a consensus of what those outside of the church would like, and he shows that there is a ton of hope, because those outside of the church are not opposed to coming to church at all, they just want it to be Christ centered in its teachings and in its actuality. What the outside would like to see more in church actually matches up perfectly to what we should be doing in the first place:

1) I wish church were not just a sermon or a lecture but a discussion.
2) I wish the church would respect my intelligence.
3) I wish the church weren't about the church building.
4) I wish church were less programmed and allowed time to think and pray.
5) I wish the church were a loving place.
6) I wish the church cared for the poor and for the environment.
7) I wish the church taught more about Jesus.[130]

This list should break our hearts and open our eyes. Is any of these seven points not exactly what Christ would want us to do anyway? How sad it is that we have apparently drifted this far away from Christ at the center; but how our eyes should now be open to it. Are you or your church going to make any changes, or simply ignore this and continue focusing on those inside the church walls of your local building? The choice is yours, but we all know what Christ would do.

Byzantine Empire "Part 2" Anyone?

We discussed the fate of the Christian Byzantine Empire in the first chapter, but re-read this summary quote from Wikipedia on the demise of this Christians Empire; once you finish reading it again, re-read it and ask yourself if this is what they will be writing about the United States and Europe as we approach the 22nd century:

> *"During its thousand-year existence the Empire remained one of the most powerful economic, cultural, and military forces in Europe, despite setbacks and territorial losses, especially during the Roman–Persian and Byzantine–Arab Wars. The Empire slipped into a long decline, with the Byzantine–Ottoman Wars culminating in the Fall of Constantinople and its remaining territories to Muslim forces in the 15th century."* [131]

Let's re-translate what this could look like in the 22nd century, if Christians continue to not take their Christ-ianity serious:

> *"During its 1500 year existence in Europe and its climatic rise in the United States, the new Christian Empire remained one of the most powerful economic, cultural, and military forces the world knew, despite setbacks and territorial losses, especially during the political correctness times, and loss of youthful members in the 21st century; followed by several Islamic jihads that crushed the remnants of Christians in the Middle East, African, then Israel. After the attempted restoration of Christ in Christianity in the 2050's, we saw the empire briefly re-establishing it's dominance, but it was too little too late as all the older generation had died*

> *away, not putting enough effort into the future; the Empire slipped into a long decline, with Europe succumbing to Islam in 2080, when France's president and parliament voted to adopt Sharia law; culminating with the United States' official emergence as a type of atheistic communist nation, in an attempt to resist the tide of Islam."*

While I readily admit that this is beyond imagination and maybe even a little comical, (I took liberty to go to the extreme), is a milder version of this that unrealistic? Let's do the math – based on current trends, (if they do not speed up nor slow down), Europe will be approximately 20% Christian, (as in believing in God and Christ as Savior), by 2030, and the United States will be less than 50% Christian by 2040. These are of course subject to change, (for better or for worse), but the trend is there and no one can deny this. Although I have admitted my previous quote may be somewhat exaggerated, many former Byzantine citizens would have said the same thing had they read the collapse of their 1,000 year old empire. To tie in the seriousness of these 3 dilemma, let me reference that I am not the only one to share this similar assessment of Christianity's future, if Christians continue their current trend of un-focused mindset.

So what is the most serious/dangerous ideology for Christianity in the 22nd Century? While I would say anything outside of Christ, I will again quote one who is much more qualified to answer such a question:

> *"Islam is willing to destroy for the sake of its ideology. I want to suggest that the choice we face is really not between religion and secular atheism, as Sam Harris, Richard Dawkins, Christopher Hitchens, and others have positioned it. Secularism simply does not have the sustaining*

> *or moral power to stop Islam. Even now, Europe is demonstrating that its secular worldview cannot stand against the onslaught of Islam and is already in demise. In the end, America's choice will be between Islam and Jesus Christ. History will prove before long the truth of this contention..."*[132]
> - *Ravi Zacharias, The End of Reason, pgs 126-27*

So according to the expert commentary of Ravi Zacharias, the only real choice is Jesus Christ. So I truly hope and pray that all camps; church members, non-church members, believers, and unbelievers, will truly ponder the "truth" of "truly" following Jesus Christ; not in some Hallmark commercial type way, but in truly following Christ. As Ravi points out and as we have discussed in this book, all issues will be resolved by simply following Christ. From the pointlessness of life outside of God, Christ making sense of all areas where other claims fall short, the dangers of fascist Islam, to hypocrisy in the church; **all** are solved by truly following Jesus Christ.

May we learn from the history of the Byzantine Empire and its fall, as well as look in the mirror and re-evaluate our own historic walk with the Lord, and our churches' history, and make adjustments to realign them with Christ instead of watching them fall apart, like the Byzantine Empire eventually did. The Byzantine Empire fell due to internal strife and fighting of Christians against Christians, coupled with the constant pressure of a fascist Islam; their main fall was taking Christ out of the center. May we re-evaluate their Christian fate with that of the United States of America. What will our history be as we enter the 22nd century and beyond?

> *"In all that we say and do we are concerned with nothing but Christ and his honor among people. Let*

no one think that we are concerned with our own cause, with a particular view of the world, a definite theology or even with the honor of the church. We are concerned with Christ and nothing else. Let Christ be Christ." [163]
- *Dietrich Bonhoeffer*

(I absolutely love that quote – **"Let Christ be Christ."** May we too, share and express this same type of sentiment – not Church, preference, or doctrine, but simply **"Let Christ be Christ."**)

Tell Me the Good News…

"He who gathers crops in summer is a wise son, but he who sleeps during harvest is a disgraceful son."
Proverbs 10:5

On a positive note – we must remember that Christ is still the answer to all, and nowhere in scripture does it say that Christendom will end in the United States. I do think it is a tragedy that we have let the Christian West deteriorate to what many are calling a "post-Christian West," but the good news is that even though we seem to be standing still, Christ's Kingdom continues to grow elsewhere. In a wonderfully written book "The Next Christendom" by one of my favorite authors, historian Phillip Jenkins; Jenkins describes that while Europe is on course to fall below 20% Christian by 2025, Latin America boasts of 511 million, Africa 389 million, and Asia at 344 million Christians.[166] In light of this, it is quite sad when I myself have received numerous emails from Nigerian believers saying "we need Bibles!" Much like KP Yohannon points out as founder of Gospel for Asia (GFA.org), America boasts of all types of Bibles and scripture messages, while the vast majority of believers do not even have one single Bible. This point is even further made when you look at our Brothers and Sisters throughout Asia. As I mentioned earlier, the persecuted church of communist China was approximately 700,000 members in 1950, and the more persecution they suffer, and jail time they serve for propagating the Gospel message and being forced to meet in secret, has resulted in their numbers today being approximately 90-100+ million. This is remarkable when we see Europe who suffers no real persecution for being Christian, ever increasingly turning their back on Him and seeing their numbers plummet to about 25% Christian today, (with the United States showing signs of following Europe if they do not act now). This should cause us to re-evaluate what is a blessing. We tend to say "we are blessed because we have so many material goods; they are not because they do not have materials and they are persecuted," but just as Christ Himself said, "blessed are those who are persecuted for my name's sake," or as Paul showed that by his and other disciple's chains (being in prison for their following of Christ), increased the faith of others, they counted their persecution as a blessing; our material advantages and extreme freedom has resulted in us being somewhat materially focused and watered down in our faith. (Christ did say, "you can not serve Him and money (mammon)."

While I see no reason to want persecution, of course, I do believe as Christ said, "He who is given much, much will be expected of," that we should be supporting such organizations as Voice of Martyrs (persecution.com), Gospel for Asia (GFA.org), Asian/African/Central and South American indigenous ministries, as well as other accredited foreign ministries of course with at least 20% of our income as well as church income, (plus another 10% for our local church and community). All of our incomes belong to God anyway and we can't take it with us, so there is no reason why we should not be reaching out while there is still time, in building the future of Christianity. I think Francis Chan sums this point up well in his bestseller, Forgotten God, where he calls for us (as I do) to seriously reflect on our true selves in light of Christ: *"Do you remember the story where Jesus saw people putting gifts into the offering box? At first some rich people gave, and it sounds like their contributions must have been monetarily large. Then Jesus pointed out a widow who put in two small copper coins. Jesus commends this woman, whom the world – those people with power and money – overlooked and perhaps even derided. Jesus praises her for her revolutionary faith, for holding nothing back. What if you could hear the voice of the Holy Spirit and He asked you to liter-*

ally give everything you owned? What if He asked you to sell all your possessions and give the money to the poor? Could you do it? Before you start explaining why He would never ask that of you, take a moment and answer the question honestly. It's not out of His character to ask for everything."[162] *(See appendix on ways to get involved in planting seeds with the American dollar why we still have adequate time)*

> *"We must fight the battle against gaining wealth as an end in and of itself instead of keeping our mind set on the fact that wealth is only a means to an end. Wealth is a highly dangerous possession. Too easily does it become a way to increase our own comfort and luxury instead of a tool to help those who suffer, and by so doing bring glory to God."*[133] *(William Wilberforce, Real Christianity pg 87)*

So while it can be sad to see ourselves living in an ever growing post-Christian nation, we should rejoice in Christ always, and once we focus on re-tuning ourselves and our churches/organizations to Christ, we can still turn the tide of the United States heading to a post-Christian nation, but DO NOT take this for granted, the United States will continue to be a less Christian nation and more secular/atheist/Islamic if we do not scrape the scales from our eyes and wake up to the realization that we have not fully:

1) Been Centered on Christ
2) Remembered our persecuted Brethren in Christ, or supported them the way we should; nor have we taken the threat of fascist Islam seriously.
3) Educated ourselves and pushed for our elected officials to teach the facts in our schools; if there are facts that point to a Creator, Designer, etc., these should be brought into the open

4) Re-tune our churches to Christ-ianity and not Church-ianity (and recognize that there is a difference)
5) Known our Bibles – we must truly know our Bibles. This does not mean from a theologian or seminary perspective, but more from the Lollards of John Wycliffe's days – these poor priests preached the basics of the Old and New Testament to all they came in contact with, (Similar to John Calvin's emphasis on the Scripture's call of the priesthood of all believers, and their roles to be just that). If we profess to be Christians, we need to at least understand what that means. Pastor Richard Wurmbrand repeatedly wrote of the importance of memorization of Scriptures in preparation of a day, when like him, we may spend 14 years in prison for following Christ.

Before this very introductory book, did you realize that persecution of Christians has ran rabid throughout Asia, Africa, and the Middle East; and still does? Did you realize that our educational systems continue to promote atheistic teachings, regardless of the evidence or lack thereof? Did you realize that our churches are on the decline, while the future of Christianity looks to be set in Latin America, Africa, and Asia? What will you do to help? What can you do? (First of all pray, but also please see the Appendix on more applicable ways you and your church can help and get involved; but if we are Christ-ians, getting involved can not be an option).

> *"As I travel, I have observed a pattern, a strange historical phenomenon of God 'moving' geographically from the Middle East, to Europe, to North America, to the developing world. My theory is this: God goes where He's wanted."* - Phillip Yancey

John Wycliffe and The Lollards

⁹But you are a chosen people, a royal priesthood, a holy nation, a people belonging to God, that you may declare the praises of him who called you out of darkness into his wonderful light. ¹⁰Once you were not a people, but now you are the people of God; once you had not received mercy, but now you have received mercy.
1 Peter 2:9-10

NO this isn't some attempt to say "let's follow some 14th century group today and start a new denomination!" I am pretty much using the Lollards to hopefully promote the exact opposite of this! Before I can really use their example to make my point however, I will need to at least give you a very brief run-down of who they were and how they faced many of the similar problems we are facing today, (but to a much greater extent).

In a nutshell – the Lollards were really more of a movement founded by John Wycliffe (c.1330-1384), who was an Oxford professor, philosopher/theologian, who developed a number of doctrines and is considered by most way ahead of his time and the "Morning Star" of the reformation to go back to the bible as supreme authority and not the Roman Catholic Church. He taught that all persons had the right to have the bible in their own language and was thus charged with heresy a number of times. Among his greatest contributions to English literature was his inspiration of the translation of the Bible into Middle English, the first complete translation in the language, and a notable influence on the English language itself.

Wycliffe was brought to trial in 1377, and he and his doctrines were formally condemned in 1382 by Pope Gregory XI, who ordered that he be arrested, but his order was never carried out; until finally in 1382 the Archbishop of Canterbury condemned him and his writings, but Wycliffe himself was undisturbed and continued to write until his death in 1384. Wycliffe was finally condemned 41 years after his death: his books were burned and his body was exhumed and burned, with the ashes scattered.

One of the practical initiatives undertaken by Wycliffe was the training and commissioning of laymen, or poor preachers, whose task was to teach the Scriptures throughout all lands in accordance to Christ's great commission. Wycliffe stressed that his intentions were not to start some new movement or to replace the church, but to fill the growing gaps by the established church. Their simple focus was to go in groups of two, to offer Christ-like assistance and tell the core of messages found in the Gospels. Many of these poor preachers carried with them hand written copies of the Gospels in the common English language of the day instead of the Latin, so that all could understand. It is interesting to note that the printing press would not be invented for about another 100 years, so all of these Gospel tracts had to be hand written, (We can order 1,000 copies of the Gospel of John in almost any language for a reasonable price today). These poor priests followed the doctrine of the "priesthood of all believers," and saw it as our duty to truly imitate Christ as well as tell others what Jesus' words are vs. the churches, (which often differ significantly).

The years after Wycliffe's body was dug up and burnt, the Lollards faced increased persecution by both secular and ecclesiastical authorities to stamp out the movement in its entirety. The persecution did gradually become less severe and it became clear that the Lollards would survive as an underground movement. A movement focused on telling all who would listen about the Gospel message, to build simple house type churches, and teach the common people, (who were being kept in the dark), about the simplicity and truth found in Christ. In the 1450s, during a lull in action against them, Lollards began again to evangelize and plant new groups. The reading circles were still influential means of attracting new adherents, and the authorities were unsuccessful in their efforts to restrict the production and distribution of Lollard literature and vernacular versions of the Bible. Lollard beliefs were passed down within families and through trade contacts. Sermons were written down and distributed to adherents and to interested inquirers. Lollard schools were also operating to instruct members of the movement and to prepare them to "reach the world" with Christ's message; to follow God rather than man.

Christian History Magazine, (volume II, No 2, Issue 3, John Wycliffe and the 600th Anniversary of the Translation of the Bible into English, Christian History Institute 1983, pg 17), summarizes it well:

"Poor Preachers –

Probably inspired by St. Francis and his street evangelists or by Luke 10:1-4, Wycliffe sent out from Oxford his order of 'poor priests' or 'itinerant preachers,' who traveled the countryside, lived and dressed simply, and preached wherever people would listen. As early as 1377, they were denouncing the abuses of the church, proclaiming the rediscovered understanding of the doctrine of the Eucharist, and teaching biblical thinking from which would come right living. Many were graduates or undergraduates, probably faithful students or colleagues of Wycliffe. Most were ordained, but not tied to a parish, free to travel. Later Wycliffe employed committed laymen, (simple people that had a love for Christ and telling others of His message). He defended their right to preach as long as they had accepted the divine call. He called them 'evangelical men' or 'apostolic men.' It was for theses 'poor preachers' – both lay and ordained – that Wycliffe prepared his tracts, his skeletons of sermons, and his paraphrases of the Bible – all in the English dialect of the people."

The actual term "Lollard,"(Lollardi or *Loller),* was the popular derogatory nickname given to those without an academic background, educated if at all only in English, who were reputed to follow the teachings of John Wycliffe in particular, and were certainly considerably energized by the translation of the Bible into the English language. By the mid-15th century the term *lollard* had come to mean a 'heretic' in general. The alternative, *Wycliffite*, is generally accepted to be a more neutral term covering those of similar opinions, but having an academic background.

Although Lollardy can be said to have originated from interest in the writings of John Wycliffe, the Lollards had no central belief system and no official doctrine. Likewise, being a decentralized movement, Lollardy neither had nor proposed any singular authority. The movement associated itself with many different ideas, but individual Lollards did not necessarily have to agree with every tenet. Fundamentally, Lollards were anticlerical, meaning that they disapproved of the corrupt nature of the Western Church and the belief in divine appointment of Church leaders. Believing the Church to be perverted in many ways, the Lollards looked to Scripture as the basis for their religious ideas. To provide an authority for religion outside of the Church, Lollards began the movement towards a translation of the bible into the vernacular which enabled more of the English peasantry to read the Bible. The Lollards stated that the Church had been corrupted by temporal matters and that its claim to be the true church was not justified by its heredity. Lavish church fixtures were seen as an excess; they believed effort should be placed on helping the needy and preaching rather than working on lavish decoration. Icons were also seen as dangerous since many seemed to worship the icon rather than God, leading to idolatry. Believing in a lay priesthood, the Lollards challenged the Church's ability to invest or deny the divine authority to make a man a priest. Denying any special authority to the priesthood, Lollards thought confession unnecessary since a priest did not have any special power to forgive sins. Believing that more attention should be given to the message in the scriptures rather than to ceremony and just "talking."

Lollards also followed similarly to Gandhi's (and moreover to Jesus') approach of non-violence, as they were for the most part against war, violence, and abortion. Lollards were effectively absorbed into Protestantism during the English Reformation, in which Lollardy played a role. Since Lollardy had been underground for more than a hundred years, the extent of Lollardy and its ideas at the time of the Reformation is uncertain and a point of debate.

So what we conclude with the Lollards is:

- They were persecuted for following Christ; had to operate underground (much like most persecuted churches to in other countries do today)
- Oxford University eventually had to expel Wycliffe for not following the "status quo" – similar to what we will face (see the Expelled documentary), in an increasingly hostile academia towards the God hypothesis and moreover to Christ.
- They used basic apologetics.
- They focused on the priest hood of all believers, and our responsibility and privilege thereof

- Simplicity of serving the poor vs. lavish and costly church buildings; formed house type churches and cell groups
- Follow Christ and Scripture as opposed to the governing church and authorities
- Following the Book of Acts – went out 2X2 to propagate the word to anyone who would listen.
- This was all happening during a time in history when Islam was attacking the remains of a weakened Byzantium Empire and thus placing pressure on southern Europe of possible invasion; coupled with the western church continuing to fight its eastern counterpart (resulting in Christ-ian vs. Christ-ian and fascist Islam vs. both, (similar to today and the future)).
- The Byzantine Empire would collapse with the fall of Constantinople in 1453 to Islam – during which many Lollard groups were underground due to persecution by the church.
- A small group like the Bolsheviks or Nazis were able to throw their dominantly Christian countries into an atheistic mindset; Islam wiped out the majority of Christian communities throughout the Middle East; Faith survived in these communities from small dedicated believers forming groups like the Lollards.

So although this movement was over 600 years ago, I think we can see how very badly and urgently a similar movement is needed today. This movement basically knew more on the subjects I have covered in this book 600 years ago, than we do today. They knew what it meant to have to choose between death or Christ, they new how real the threat of fascist Islam was, (it had only been driven out of France a few hundred years earlier and was threatening European invasion yet again); they new more about the simplicity of being able to hold church in a home instead of a 30 year mortgage building that handicaps the churches of today; they new more about why we don't need to go to a "temple" to worship anymore, but instead that we take the temple (of the body) to the church; they new more about non-violent fighting – how to protest without using a sword (although different, similar to the approach made by Gandhi in the 20th century); all in all, we truly need a "Lollard" movement or mindset today.

Think about it – a group of friends begin meeting with others at a local church, community building, Starbucks, bookstore, or home, and along with scriptural readings they become active in educating themselves through prayer and study and going out 2X2 to college campuses, shopping centers, door to door, etc, finding like-minded people and growing in Christ. What could be more Christ-centered than this? As I have mentioned, I don't agree with Mormon doctrines for the most part, but my Mormon friend has shared many experiences where he says the Mormons go 2X2 to different homes to knock on the door and offer to study and fellowship. Many say "no," but there are many who are alone and need companionship and fellowship and now they have an offer at the door! Needless to say, the Mormon Church is growing much faster than most other churches simply because they're following this Biblical doctrine, (although they are not using the Bible solely unfortunately). Why is it when I see two people knocking on my door with a Bible in hand I know it is either Jehovah's Witnesses or Mormons? Why not any other denomination? Why not a "non" denomination or simply a Christian Brother or Sister offering me a free Bible and an invitation to their church or group? Great question in my opinion...

I feel that this type of "Lollard" mindset is crucially important now and for our future. If we at least have a broad understanding of persecution, apologetics, and authentic

Christianity, we will be able to lovingly deal with any objections and share the message of the Gospels to a large audience, just like Christ instructed us to. For my part, I am willing to help you start this type movement in your own community (Lord willing), by giving you a free ministry start-up kit that should really help you get started in a big way. This is for those serious about their faith of course, but if you really are serious about it, please see appendix F at the end of the book, or visit www.TheLollards.org

I hope you take advantage of this and get involved. But whether you do or don't, the important thing is to utilize the content of this book to come up with a cumulative case for equipping yourself to understand these precepts that are becoming increasingly important today and more so in the near future. Call it "Lollards" or whatever you would like, we desperately need genuine disciples (not scholars) to take up the challenge of the "priesthood of all believers" and place Christ at the very center of their being and lives.

John Chrysostom (349-407) lived in a world much like ours is today – full of tensions, social injustices, love of money and material possessions, and a "me first" mentality guiding every decision. Similar to the Lollards 1,000 years later, he admonished us to root our work in the gospel as revealed by Christ in Scriptures. He continually emphasized the church's role as well as our role in keep our church buildings simple and using the money to support those who are poor and in need; to worship God in imitating Christ to our fallen, selfish, and needy world today. (Picture taken by author's wife at St. Patrick's Cathedral New York)

What about you?

"Good News: 1/3 of Americans read their Bible at least once a week. Bad News: 54% can't name the authors of the Gospels. 63% don't know what a Gospel is. 58% can't name 5 of the 10 commandments. 10% think Joan of Arc was Noah's wife."
- New York Times, 12/7/97

While I have continually recognized that I am a novice at best, nothing in this book should really be taken as controversial, because chapter 1 is taken straight from history, chapter 2 from a macro level of secular science and philosophy that no one really disputes, and chapter 3 is mostly straight from the most obvious Biblical passages as well as statistical analysis of how modern viewers see the church vs. Christ. One might say what qualifies you to write this book; which I would have to ask, what doesn't qualify me or anyone else to write a factual book from the heart? While I readily admit that I have a plethora of my own problems, we need to recognize that we're all hopefully on the same side; that of being in Christ. Almost every religion of the world, including atheism, recognizes Christ as "truth," but they differ on all fronts when it comes to how we and the church mess up the interpretation of what that "truth" is and what direction we go from there. But if we just take Christ as He is, and place Him at the center of our lives, no one can deny the positive effects that will play. (Imagine if everyone truly had "Christ in them" and treated everyone around them as Christ would, and so on and so forth. Would there be any hate or conflict?) Please consider what I have written, and please by all means go to Amazon or my website to tell me if I did or said anything wrong or that Christ would not agree to, so that I too may learn and grow in this wonderful person – this wonderful Savior of us all.

I will end with three quick personal stories that are not exactly "perfect stories," they're real stories that have probably taught me just as much about my Christianity as I could have them. I of course believe the Lord showed me these three different and wonderful people to help with my own discipleship, and give me a lesson on what being real is, and what it means to Christ being in you. I told you about my atheist friend Jenna who I worked with in the microbiology lab at the University of Arkansas, to whom I did not give satisfactory answers for why there must be a God, (see chapter 2), and that while we were good friends, she moved away and I have not seen here since. I hope that my years of friendship at least gave her somewhat of a seed that she can reflect on the person of Christ, but hardly a success story on "how I reached her," etc. The Lord somewhat used our encounter to stick with me more than 10 years later, to show me to be prepared next time, but more importantly to always show that patience, love, and compassion that Christ showed all. More recently I met one person when the whole "MySpace" thing was popular. She was in a Nazi group and a stripper from the Virginia area named Erika. While Erika had a somewhat rough childhood, she still had the yearning for God in her life, but she had tons of questions, (and still does). While I would have to classify myself as an "average joe," we all have our part to play. So I mailed her a very detailed Children's Bible that went thru the Old Testament and New in about 500 pages. It was a very basic paraphrased Bible that is much easier to understand. I mailed it to her along with about 10 different biblical movies to help her better understand the Scriptures. I spent about the next year and a half going through this Bible and the movies, and it was amazing to see how she would get almost mad at some points, but her hunger for God and the truth would keep her coming back continually. We finished the Bible and have now started on a real version of the Bible. She has given herself to Christ and accepted Him fully, although she still is not overly crazy about the church. She has tried several different ones, but she said the church always looks at her funny because she has bleached blonde hair that kind of sticks out from the "norm." She has also quit stripping and is working at an animal shelter at present. Although she has had some depressing times having turned away from some of the party life of drugs, stripping, etc, she says it is all so worth it for the wholeness of what Christ gives her. She is now actually reaching out to some of the Jew-hating Nazi organizations with which she used to be affiliated, with the story of Christ, as well as some of her stripper friends. While the Lord has allowed me to learn probably just as much from Erika as she has from me, we have both learned more about Christ and His love. Many would of course say, "Why are you affiliated with her!?" When I ask myself, "What would Jesus do," I think we would all agree that He would spend as much time as needed with her and anyone else who seeks Him. Here is one of my and Erika's (aka – princess) first and last emails (note the dates):

Nov 24, 2007
Subject:
From: Princess
Body:

No I would not like to follow the same path as any filthy Jew of course. I want to be who I feel I am, and that is what I do. I will not follow in Jesus the Jew or any other Jew. Also, there are things that I have no control over, that have a hold of me, which I can let go of. To give up my actions, would also be giving up on myself, it is who I am

I was hoping to avoid this with you, you obviously know and understand more about such things than I do. But I do know the important thing in my mind: Was Jesus a Jew? Of course, Jesus was a Jew. He was born of a Jewish mother, in Galilee, a Jewish part of the world. All of his friends, associates, colleagues, disciples, all of them were Jews. He regularly worshipped in synagogues.

Really do not wish to be rude to you or criticize you at all and it may look that way, but how could I even consider a thing like this?

Have a great day

——————— Original Message ————————
From: (Author) aka – Alliance100.com
Date: Nov 25, 2007 2:42 AM

Thank you for looking; I will continue to pray for you my friend. Have a wonderful day & God Bless. *Feel free to ask me if you have any questions. **I hope you have time to view the video in my bulletin that I sent you. Thanks again ;o)

——————— Original Message ————————
From: Princess
Date: Nov 26, 2007 10:28 PM

I told you I would look and I have looked. You would not be interested in my thoughts. I am way too far away from that to even consider such a thing.

←Almost 2 years later with study, patience, and prayer:→

Aug 3, 2009 8:23 PM
Subject:
RE: Hi Mr Alliance ☺
Body:

Hi Mr Alliance! How are you? I've been meaning to tell you... I'm VERY proud of you too!! Let me explain why.... lets just say that I would not have known you or you had not have helped me so much like you did.... and gave me advice when I asked for it and such... more than likely I would have seen and observed how the majority of other Christians out here are. I would have thought.. its okay to have occasional sex if nobody knows... its okay to act one way at church and another with other friends... its okay to go party if its only every once in a while or if no one knows will find out.. and so on...

I wouldn't have known any better. That's definitely what I would have seen from others...i really would have thought that acting like that is acceptable behavior.. and its not!

And im SO proud for living the way a Christian really should be!! That is something I just haven't seen in very many people at all since I have started this way ☹

However.. one last thing ;) : Im not really sure what it is we actually have to go over any more... we have finished the bible and I mean I have started it again.... but is that something that you really want to go through with me again...? It was so much work for you too, when I can just do it on my own now. What do you think? ;)

As far as the church thing goes, to be honest I have not been since the last time I told you I did. I have been busy looking for a job... getting adjusted...trying to help my friends the best I can... Not an excuse either I know. I need church... I need to go badly, finding a good one is hard I guess though.

I am doing great Mr Alliance ☺ I am the happiest ive ever been in my life! I never even knew that there could be a happiness like this.. I guess ive never been I happy before! This is the best thing in the world!! I love Jesus so MUCH Mr. Alliance ☺ Even though I am not accustomed to my friends testing me (and it doesn't make me too happy, lol)... that was in fact exactly what I needed... to be on my own. I trusted in Jesus, just like I remember us talking about... and it was right! Thank you for everything Mr Alliance ☺

———————————**End of emails**

I still am being blessed working with Erika, and she has taught me just as much as I ever could her. She is still looking for a church that doesn't look at her funny, judge her for her past, or seem fake. Please pray that she finds one that places Christ and His love at the center of their worship; and please pray that next time we see a "Nazi/former stripper/drug user," that we will not give up on them, but show them the love that Christ would.

Another friend, Matt, is what I would probably call an agnostic, (he doesn't know if God exists of not). He is also what I would call a bit of a philosopher. He is quite intelligent, and he is very skeptical of organized religion. I met Matt at work and we began just asking questions and answers back and forth. He had a lot of difficult questions on the existence of God, that hopefully I at least answered somewhat satisfactorily. He does admit that the evidence for some type of God is at least likely. He also asked me several questions on *"Why are Christians so arrogant, hate gays, and everything?"* (He has had a bad taste left in his mouth by church/Christians) I again, hope I answered these satisfactorily and I did apologize on behalf of people whom he had experienced un-Christ like things from. Also, when Matt and his atheist wife asked me to help them move one evening, I showed up with my truck to help. Many of his friends did not show up like he thought they would, and once we were done, he really showed a lot of gratitude and his wife then accepted me as friend on "MySpace."

So I have known Matt for a little over one year; what is his status? He is still agnostic and his wife atheist to my knowledge. But the point is that my purpose as a Christian is to love them and to be there anytime they need. This is what Jesus taught, is it not? My goal honestly is to hopefully remove some of the intellectual barriers that he has against the Gospels, and then help him in the direction of being a Christ-following agnostic. That's it. I am nobody, and can therefore not "convert" (as so many like to poignantly call what they attempt to do), anyone. Only the Lord's Spirit can convert one to be a full follower of Him; if I am able to help one first follow Christ alone, (even if they don't want to accept everything else with it), then I think I have done my calling, as Paul pointed out: "I planted, Apollo watered, the Holy Spirit grew." Sadly, many Church-ians think that you have to do all three parts yourself. (Often referred to as EGO, (Edging God Out)...

Another good friend of mine is an Egyptian supplier that I work with named Nouran. She is a very kind and sincere Muslim. While I do not agree with the vast majority of Islamic beliefs, that does nothing to change my love for her or her family. I have known Nouran a little over one year, and we have had some in-depth talks about religion and Jesus. It is ironic, but she too has concluded that many of the religious dogmas at her mosques take away from God and spirituality. We must reach people where they are, (just as the apostles did), and once again, my hope is to help her see the "truth" of becoming a Christ following Muslim. Is there any better place to start than with one following Christ? (I hope that surely no one would argue with me on this point).

After talking sincerely for a few months, I asked her if she would like to watch a movie that is the word for word message of the Gospel of John. She said she would, so I sent it to her. She contacted me afterwards saying how the portrayal really touched her, and that it did make Him seem more than a mere prophet. While stoning is still practiced in many Islamic countries, I now can use this video I gave her as a springboard for discussion; I have often asked her if she thinks the following of Islamic laws on stoning or Jesus' way of "Whoever is without sin may cast the first stone," seems the most Godly. She has so far always came back to say Jesus' way. So where is she now? Has she accepted the "stamp" of Christian and settled down in some generic church? No. She is a Muslim, but she is increasingly considering herself a Christ-following Muslim. (She said she is going to watch the movie again carefully and take notes, which will be good for both her and me.)

My other best friend of childhood is a Mormon now. I have just given him an audio copy of the New Testament which is acted out, to help in his walk. We have been meeting semi-regularly to study and he has asked me if I will read and study some more about Mormonism, and I have agreed, (although I disagree with most of their doctrines). While others I hear yell and scream on the radio that "he is not Christian," I must ask, are you? I trust the Lord fully, and if Matt asks me to watch some atheist tapes, Nouran to read up on Islam, Scott on Mormonism, and Erika asking me questions regarding her past life, then as a follower of Christ, I should be willing to meet them where they are, and understand the perspective they're coming from. Is this so wrong? If not, then why do our churches and many Christians take the approach of "stay away from them?" I don't agree with most of Roman Catholicism, for example, because it like Mormonism is not fully in-line with Scriptures, but if a Catholic friend asks me for help, then I will do my best to tell them about Christ, and keep the focus on Him. If I do that, I have confidence that He will take care of the rest. Can anyone reading this honestly say this is not what Christ would do? Can anyone say that placing Christ at the center is wrong?

Dietrich Bonhoeffer who was martyred by Hitler, summed this problem up well in his wonderful book **"The Cost of Discipleship"**:

> *"Cheap Grace is the deadly enemy of our Church. We are fighting today for costly grace. Cheap grace means grace as a doctrine, a principle, a system. The word of cheap grace has been the ruin of more Christians than any commandment of works. Instead of opening the way to Christ, it has closed it. Instead of calling us to follow Christ, it has hardened us in our disobedience. We are no longer sure that we are members of a Church which follows its Lord. We must therefore attempt to recover a true understanding of the mutual relation between grace and discipleship. The issue can not longer be evaded. It is becoming clearer every day that the most urgent problem besetting our Church is this: How can we live the Christian life in the modern world?"* [134]

My humble, sincere, and urgent prayer is for us to all repent as the people did in Nineveh following Jonah's warning for them to fall on their faces in repentance and worship to the Lord; to take our following of Christ to heart, and become true followers of Christ over doctrine and dogmas. Christians do have a history of mistakes where we have become disconnected from Christ, but still called ourselves Christians; we need to ask for forgiveness for these mistakes, repent (which means quit not following Christ, and to truly begin following

Him in this instance), and change the misconception that many of us and our churches have caused in society that ***"I like your Christ, but not your Christians. Your Christians are so different from your Christ."*** To do this, we need to first and foremost imitate Christ in our churches, in our families, our individual lives, and most importantly in reality.

> *"I hope you will not mistake what I am saying here. I'm sure that on the day we stand in the presence of Christ, we will discover that there are some who understand these things and affirm them but whose lives show no evidence that Christ has transformed them. Authentic faith will always be evidenced by changed lives. Surely this is what Christ meant when He referred to those who take His name and yet on the last day will hear the dreadful words: 'I never knew you. Away from me, you evildoers.'"*[135]
> *(Matt 7:23)*
>
> -William Wilberforce, (pivotal figure to the abolition of the slave trade), Real Christianity, pg 59-

Conclusions –

> *"For the Christian movement, the founder must be able to be seen in the lives of the found. People observing us ought to be able to discern the elements of Jesus' way in our ways. If they cannot find authentic signals of the historical Jesus through the life of his people, then as far as we are concerned they have the full right to question our legitimacy."* [136]

I hope and pray that this simple book from a novice "wanna be" Christ follower, has at least addressed the issues that are not "hypothetical" but are "factual." It is such a tragedy that if we look through the history of Christianity up to today, we see so many Christians that do not follow Christ. In Gandhi's autobiography, he labeled the book "The Story of My Experiments with Truth," and in his introduction, he says, "What I want to achieve – what I have been striving and pining to achieve these thirty years – is self-realization, to see God face to face, to attain Moksha (salvation). I live and move and have my being in pursuit of this goal." [177] While Gandhi came to see the truth in Christ, he was highly discouraged by self-proclaimed Christians that drove him away from the title "Christ-ian;" it is my prayer that this simple book has helped for us to strive to be more like Christ who is the "Truth" itself, and make certain that we do not drive anyone away from Him, as Gandhi was. The main way we can do this, is by putting Christ at the center of everything we do, (just as scripture mandates). But what does this look like in more descriptive terms? I would summarize it as the following:

1) Truly know Christ in your life. This encompasses truly knowing and having a relationship with Christ, and by default, your "light will shine before others." Christ described this Himself when He said "Love the Lord your God with all your heat, soul, and mind, and the second is as the first; love your neighbor as yourself."

2) Know why you believe what you believe. As 1 Peter 3:15 put it, "Always be ready to give an answer to anyone who asks why you have your faith." As discussed in chapter 2, this simply means being able to at least know what Christ taught, the Bible, Christ's fulfillment of prophecy, that science and history all point to a God/Christ as filling all of our questions and concerns.

3) We also need to be able to reach the culture around us. In an ever-growing pluralistic/relativistic/secular culture, we need to be able to have the knowledge and ability, (like Paul did through his ministry), to be able to answer why Christ is the only way, (in a loving explanatory way; not a rude/crude "in your face" type of approach). This will involve at least educating ourselves and those around us, to at least the basics of other world faiths. As Dan Kimball states: *"People in our churches will be meeting more people, becoming friends with them, and having conversations with them on (different faiths). So we need to provide basic ongoing training in the origins and basic beliefs of Christianity and other world faiths. I believe that among Christians who are out talking to people there is a high felt need to know more about other faiths so they can intelligently and lovingly talk to people about them."* [167] How will you talk to a Muslim if you know nothing about Islam, or how will you talk to a pluralist when they say "all roads lead to God," if you don't know why they don't?

Listen to this following skeptic's comment: *"If Christians want us not to think they are just brainwashed and that they basically have thrown out their capacity to think for themselves, then they should be able to articulate and dialogue with us about why they hold that Christianity is the one true religion. I don't want a lecture. I want to hang out and talk.*

Like we are talking now. Back and forth. And I want them to show me why they don't think that all faiths and all spiritual expressions are valid. Not just spout out a Bible verse but give me some reasons so I understand them and why they believe what they believe. I would respect them more if they did. But all I normally get is a quick Bible verse and a "Jesus is the only way" answer."[152] *Before you get too insulted by this, this is the same type of questions that Jesus and the Apostles regularly addressed.

4) Use at least a good portion of church and personal money to help support our Brothers and Sisters overseas. Whether we like it or not, Christianity will have its highest populations, growth, and 22nd Century future in Asia, Africa, and South/Central America. We are failing in the West, but we still have the better part of the 21st century to make a difference. Our churches are still quite full overall (unlike Europe), even if our average age is approaching 60 in traditional churches. We need to utilize our funds while there is still time, to concentrate "true" Christ-centered (not doctrine centered) growth in these countries, so that when we're gone, the seeds we have planted will last well into the future. Is this not what Christ would do? A few dollars will go a long way in these countries, so we need to fuel the fire of revival in these countries while we still have the time and resources. As Philip Jenkins stated so clearly in his book, The Next Christendom, pg 3, *"By 2050 only about 1/5 of the world's three billion Christians will be whites. Soon, the phrase 'a white Christian' may sound like a curious oxymoron. The era of Western Christianity has passed within our lifetimes."* So if we are truly wanting to make a difference, then there is no reason why our churches and us individually would not be using at least 20% of our incomes to foreign growth, (with another 10% to local growth and education).

5) Last, but certainly not least, we should have our own reformation to remove hypocrisy, dogmas, and

church-ianity from our own churches. We need to re-boot back to the basics, and that is "CHRIST" in the center of all that we say and do. As I have pointed out repeatedly, (and I hope that no one would disagree with), we must put Christ back in the center of Christianity, and thus truly know the Scriptures. Numbers and opinions do not lie; the youth are tired of doctrine/dogma over Christ, and I do not blame them. We all have a hunger for Christ, but if the church continues to fight amongst themselves and others not of their "denomination," etc, then we can expect the same results of decline in our churches. And once again, does anyone honestly think Christ would say that we should fight and bicker amongst each other over such topics as clothing in church, food/diet, who's the chosen #1 church, music, etc, or that He would say "follow Me?" I pray that we will get our focus back on Christ and then all five of these points will come into perfect view.

I must now conclude where I started this book. Morpheus presents Neo with the option of taking a red pill or a blue pill. The red pill will start him on his journey to discovering what the "truth" really is, and the blue pill will cause him to fall back asleep in his dream world, and he will never have to worry about the "truth." Neo chose the red pill in the movie, but more importantly, is which pill will – **blue or red?** Do you really want to go through life with the blinders on, always wondering why you have this "void" deep in your mind and heart? Jesus of Nazareth, known often as the Christ, repeatedly stated that He was the "truth" itself. Remember the story line of the matrix, when Morpheus says that he can not open any doors for Neo, all he can do is show him the door. He then states that all he is offering is a "chance" at the "truth," and if he wishes to know more of the truth and to see how deep the "rabbit hole" (in reference to Alice in Wonderland) goes, then to take the red pill. My humble hope is that we will choose to take the red pill for ourselves, and explore reality for ourselves. While every religious system or man-made social system has stated to "show" the truth, only the being Jesus the Christ, has stated that He is the complete essence and being of "truth." May we all join one another after this life is done, outside of the matrix of falsity, and within the being of this "**Truth**."

Which pill will you take?

Afterword...

"From the beginnings of His public ministry, Jesus set the record straight: 'I have come to change the world.' He had come to change people's thinking. He had come to revolutionize their paradigm – the way they saw the world and had comfortably settled into it, following their own desires, ignoring those around them who needed help, and figuring that was the way to do it because everyone else was doing pretty much the same thing.

Then came Jesus. 'You have heard it said… but I say unto you…'

We often think we have arrived spiritually because we have memorized a bunch of rules and regulations, or a list of do's and don'ts, and built them into our culture and lifestyle. If you just wear your Christian T-shirt once a week, always have your radio turned to the local Christian music station, don't do drugs or drink, go to church on Sundays and Bible study one night a week, take part in the summer mission trip, and don't have sex until you are married, then you are a real witness to those around you.

But Jesus came and shattered man's perception of the world. He taught about a kingdom that was more powerful than man could ever comprehend. This

kingdom, however, didn't always make common sense. He taught that the kingdom of heaven is more than clothes, music, Bible studies, and what you do on the outside. The kingdom of heaven is spiritual. The kingdom of heaven is peace. The kingdom of heaven is the narrow road of following Jesus day by day. You can't be like everyone else and follow Him. His sheep hear His voice and chase after Him; they don't just roam with the rest of the herd and think that is going to save them.
The path He has for you is very different, very special; it is unique in all of humankind. He came to change the world. How can those who truly follow Him have any less of a goal in life?"[137]
By DC Talk – Jesus Freaks, Revolutionaries, page 4

Dietrich Bonhoeffer, (who I have quoted throughout this book), was faced with a struggle between Nazism and Christianity; in his day the majority of church leaders decided to play it safe and side with Nazism; (many church leaders even took oaths to Hitler). Similar to Dietrich's day, we too will ultimately have to choose between the politically correct stances portrayed in this book or a true/non-watered down Christ. We are seeing an ever increasing-number of churches agreeing to take the path of least resistance as they did towards Nazism in Dietrich's day; too worried about losing tax-exempt status, or appearing to be "too" fundamental by the media, they are choosing to compromise and conform. A great example was when the Arch-Bishop of Canterbury said that he felt Christianity would have to adopt at least some Islamic features and/or doctrines in its future. This is coming from the head of the Anglican communion and principle leader of the Church of England?! This shows yet another example that even the leaders of the church are more worried about political correctness than they are Christ. As mentioned repeatedly, this type of compromise will determine the fate of Christianity in England into the 22nd century. As Christ Himself said, *"you can not serve two masters;"* we too must decide whom we will serve today. These decisions will ultimately affect Christianity today and well into the 22nd century. We have an uphill spiritual fight before us, but as long as we keep Christ at our center, then we have nothing to fear. The three things I would like to humbly reiterate are:

1. **We must remember our persecuted brethren in Christ throughout the world, and support them in every way possible**; while also standing firmly against such militant ideologies as the fascist version of Islam, that seeks to Islaminize the modern world. – Appendix A
2. **We must follow the "truth" of Christ**, and while that doesn't necessarily mean "Bibles in the classroom," it does mean that our educational systems should be able to teach the evidence where it leads, even if this points to a supernatural origin as mentioned in chapter 2. Our Children, young adults, and future leaders of our world deserve to at least have all the evidence at their disposal in our educational systems, (not only naturalistic/atheistic ones). – Appendix B
3. **We must re-think "the church" as the active body of Christ, (not just a church building)**; which means taking the "church" to the world, not just waiting for the world to come to the "church." We must BE Christians, not just say we are, so that our light may be seen by all as authentic. –Appendix C

Ëcumenism – (not "pluralism") – mainly refers to initiatives aimed at greater religious unity or cooperation focused on Christ:
1 Corinthians 12

[12]The body is a unit, though it is made up of many parts; and though all its parts are many, they form one body. So it is with Christ. [13]For we were all baptized by[a] one Spirit into one body—whether Jews or Greeks, slave or free—and we were all given the one Spirit to drink.

[14]Now the body is not made up of one part but of many. [15]If the foot should say, "Because I am not a hand, I do not belong to the body," it would not for that reason cease to be part of the body. [16]And if the ear should say, "Because I am not an eye, I do not belong to the body," it would not for that reason cease to be part of the body. [17]If the whole body were an eye, where would the sense of hearing be? If the whole body were an ear, where would the sense of smell be? [18]But in fact God has arranged the parts in the body, every one of them, just as he wanted them to be. [19]If they were all one part, where would the body be? [20]As it is, there are many parts, but one body.

[21]The eye cannot say to the hand, "I don't need you!" And the head cannot say to the feet, "I don't need you!" [22]On the contrary, those parts of the body that seem to be weaker are indispensable, [23]and the parts that we think are less honorable we treat with special honor. And the parts that are unpresentable are treated with special modesty, [24]while our presentable parts need no special treatment. But God has combined the members of the body and has given greater honor to the parts that lacked it, [25]so that there should be no division in the body, but that its parts should have equal concern for each other. [26]If one part suffers, every part suffers with it; if one part is honored, every part rejoices with it.

In my opinion these three areas are key to discipleship, (as outlined in the three chapters of this book), though they are often neglected:

- **Christian Apologetics** – Being able to give a reason for the Faith that you have – 1 Peter 3:15 (www.IntelligentWonders.com)
- **Christian Persecution** – To remember that the Church in the North, South, East, and West are all our Brothers/Sisters in Christ, and to help them in anyway we can, as we learn from each other. – Hebrews 13:3 (www.MartyrsCry.org)
- **Christian Discipleship & Authentic Faith** – To go to all nations telling them of Christ as one body. – 1 Corinthians 12 and Matthew 28:18 (www.TheLollards.org)

Authentic Faith

CHRIST

Persecution   Apologetics

So once again and in the same context the book began – make sure we do not miss our "train" or an opportunity to help clear the road for the future generations leading into the 22nd century and beyond. The Book of Acts is still being carried on today; may we faithfully contribute to playing our part and writing our pages, as we await Christ's return.

I hope this very simple book has at least given you a basic outline of these three points, as well as give you some insight on what the 22nd century looks like at this point in time. We can change it for better or for worse, that part is up to us; I just pray that Christ is truly in each one of us. If He is, then all of the above will come in place exactly how it should be.

Thank you and God bless –

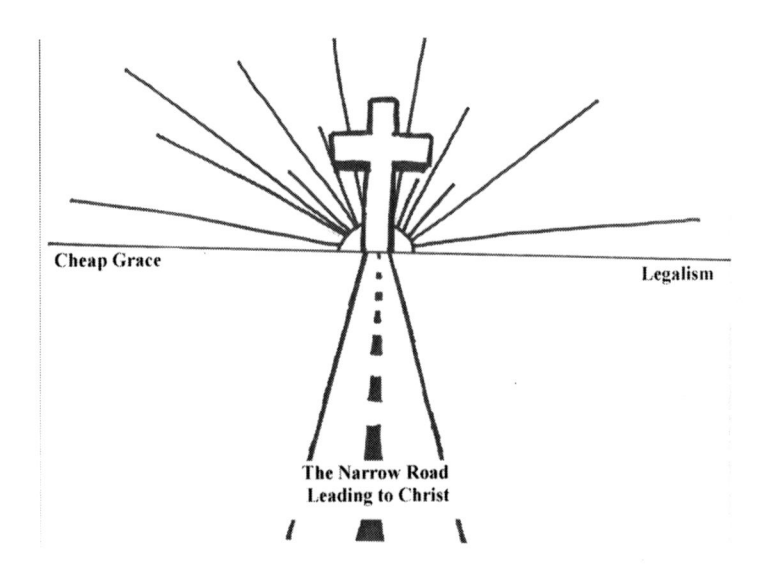

Cheap Grace   **Legalism**

The Narrow Road Leading to Christ

[13]***"Enter through the narrow gate. For wide is the gate and broad is the road that leads to destruction, and many enter through it.*** [14]***But small is the gate and narrow the road that leads to life, and only a few find it.***
Matthew 7:13-14

Friend and Pastor Marc Lien once outlined the road like this. On the far right you have the legalist, (similar to the Pharisee of Jesus' Day), who prided themselves on never missing church, keeping all the doctrinal rules, but at the same time were missing the entire picture. On the left side, you get almost a version of pluralism/relativism with the title "Christian." This could be the very liberal version of "saved by grace so therefore we can sin as much as we want." In other words, they do not really accept "grace," they are accepting sin and calling it "grace/Christianity." Both are far off from the center path of Authentic Faith that leads to Christ.

Appendices

Appendix – A

Persecution yesterday, today, and tomorrow –

Having already covered the basics of Christian persecution in chapter 1, I feel it is crucial to leave you with at least some ideas for what you can do with this information. First, continue to pray for these believers, and all those who suffer horrible human rights abuses. We can learn so much from the persecuted church about being true Disciples of Christ. It is easy to get luke-warm and turn into a "Laodicea" type of church or believer; but seeing the courage of our ancient and present Brothers and Sisters who are dying before they will deny Christ should make us really look hard in the mirror and ask if we have that same strength of faith. Let's be honest, with political correctness already plaguing Europe and the United States, it is only a matter of time before more and more "compromise" will be demanded of us. This compromise could take many forms of course.

We already have seen legislation passed in Canada that outlaws reading of some scripture that is deemed "offensive/hate" oriented. We have seen our educational system consistently attacked to push not only Christianity out, but even creation/intelligent design, no matter the evidence. While this is not like having someone put a knife to our throats, it is persecution all the same. And one big factor that I hope Christians, leaders, and churches are ready for is when the time comes to decide between compromising the Bible and our beliefs or losing our "tax-exempt" status. What do you think Christ would say about "tax-exemption?" He would probably say "who cares?" But let's be honest, if/when this time comes, many churches will "compromise" in order to keep their status. I hope I am wrong, but we will see. Either way, now is the time to educate ourselves as to what persecution means, and learn from our persecuted Brethren – who can teach us much in this area. We must be prepared to give up everything in order to follow Christ. The Scriptures are very clear on this, and so is Jesus.

So my first suggestion would be for you (right now), to go to www.persecution.com and get a free copy of Richard Wurmbrand's book *Tortured for Christ*, and sign up for their monthly newsletter to educate yourself and those around you. (You can also request this from me at www.MartyrsCry.org). Second, ask either your pastor, elder, or just have someone over to your house to give a presentation, free literature, videos, etc, on the persecuted church; you can click the "Request a Speaker" option on this website and an area representative will come to your church or small group to present. Third, you should think about being an area representative yourself. There is a real need for this field and Voice of the Martyrs could definitely use more. I am an area representative for the southern Missouri and NW Arkansas area, and I think we have about 10 total for Missouri and just one for Arkansas. (You will be given basic course material going over the history of persecution, talk to a regional representative, do a practice presentation and you're ready to have your own ministry that VOM will provide resource materials for you to distribute.) VOM has also sponsored a persecuted pastors program, Bibles for those countries that are not allowed to receive Bibles, Action packs that you and your family/friends put together yourselves to send to the needy all over the world, and they cost us literally nothing, but support a family and help build the future church, (especially if the United States does continue on its current course, where it will be less than half professing Christians by 2035).

Learn from our Brethren and tell others about their plight. Get your own group started, I have found that I can rent an entire auditorium at a college for as little as $35 per day when I tell them this is non-profit. It is amazing how these stories do not make the news, because they are deemed less or non-important than who some actor is now dating or not dating. Visiting sites such as the www.ACLJ.org, getting your state/governmental representative's information and either emailing them or even setting up an appointment with them is a great way to tell of your concerns for human rights and religious freedoms for all nations, and that it is worth breaking ties with countries like Saudi Arabia even if we do have to pay for more fuel, etc. Again – cheap gas or knowing Christ – which is more important and which Christ would say is more important? Seriously. Saudi Arabia, for example, has a strict form of Islamic Sharia Law where if one changes from Islam to Christianity, they can be executed by the government. The Saudi King has said there will never be a church on Saudi soil; and yet, they are our ally? Tell me we're not compromising already. The President of Iran proclaimed in his inaugural speech as president, that he would "destroy Christianity in Iran;" and in September 2008, the Iranian parliament passed a resolution making anyone changing their faith from Islam to another religion not just illegal, but punishable by death. Does this make you upset? It should. As Paul said: "Be angry, but do not sin." As I mention in Appendix C, Islam is in desperate need for reform, and allow its members to choose to leave the religion if they wish. What can you do? Again – write your officials, stay informed, get your churches involved, get plugged into to sites such as Voice of the Martyrs or International Christian Concern and sign up for their newsletters. (www.persecution.org (International Christian Concern has links and petitions that you and on which your churches can sign directly on-line). I would also highly encourage each of you to visit Lebanese Christian Brigitte Gabriel's site at www.ACTforAmerica.org and read her account of being persecuted and eventually forced to leave her country for being a Christian as fascist Islamists invaded. Brigitte has worked hard to get a grass-roots organization started that will help address some of these issues that are growing at an alarming rate here in the United States; I would highly recommend you consider starting your own ACT chapter in your home town, (www.ACTforAmerica.org for more information).

Open Doors ministry to the persecuted church has just released their top 50 countries for Christian persecution – can you name any? There are over 50 countries that either the government outlaws any form of Christianity or is hostile to Christianity. We must educate ourselves and others to the plight of our fellow Brethren in Christ. Gospel for Asia (www.GFA.org) has a wonderful ministry on equipping the local churches with the means and training they need to tell remote villages about Christ for the very first time; we can take part in sponsoring these missionaries. The Back to Jerusalem movement from China has plainly said they know suffering and persecution, and they will spread the word to the Islamic Middle East no matter the cost. They need our help and prayers, Brothers and Sisters. I have not formatted the above to be in any particular "poetic" order, but it is from the heart. I will say no more, but please consider sponsoring and supporting these wonderful endeavors, get your churches involved, and let us assist our Brethren throughout the world. We have the finances and ability to act today. Christ said "those who are given much, much will be expected of." Let us not slack in this commitment as we are included in those who are given much. Let's not be happy with giving 5-10%, but consider giving 20-30% of our salaries to such endeavors. We're only here for a blink of an eye and we can not take any of our money with us, so why not? What do you think Christ would say?

I have said enough – but I do prayerfully ask that you do your own searches, visit the below sites, and get involved. Please let me know if I can be of any assistance.

www.MartyrsCry.org
www.persecution.com
www.GFA.org
www.backtojerusalem.com
www.AsiaHarvest.org
www.ForTheLoveOfMuslims.com
www.opendoorsusa.org
www.ChinaAid.org
www.ACTforAmerica.org
www.persecution.org
www.christianpersecution.info
www.christianmonitor.org
www.jihadwatch.org
www.operationnehemiah.org
www.barnabasfund.org
http://www.ifcj.org

Appendix – B

What is apologetics, and what can I do?

I was asked to address our church on something dear to my heart, so I decided to talk about the importance of Christian Apologetics. Then I was quickly asked by the vast majority: "What is 'apologetics???'" - So let's start from there – the word "apologetics" comes from the Greek word "apologia," pronounced, "ap-ol-og-ee'-ah." It means, "a verbal defense." It is used eight times in the New Testament: Acts 22:1; 25:16; 1 Cor. 9:3; 2 Cor. 7:11; Phil. 1;7; 2 Tim. 4:16, and 1 Pet. 3:15. But it is the last verse that is most commonly associated with Christian apologetics.

"....but sanctify Christ as Lord in your hearts, always being ready to make a defense to everyone who asks you to give an account for the hope that is in you, yet with gentleness and reverence" (1 Pet. 3:15, NASB).

Therefore, Christian apologetics is that branch of Christianity that deals with answering any and all critics who oppose or question the revelation of God in Christ and the Bible. It can include studying such subjects as biblical manuscript transmission, philosophy, biology, mathematics, evolution, and logic. But it can also consist of simply giving an answer to a question about Jesus or a bible passage. The latter case is by far the most common and you don't have to read a ton of books to do that. (carm.org)

So now that we have gotten that out of the way, "What can we do?" – Could best be answered in a number of ways. First of all we must have authentic faith that can be seen and we also really need to know our Bibles. As we have already mentioned, the vast majority of Christians can not name the authors of the Gospels; in addition, they cannot give a concise summary of the Bible (Old or New), or even name half of the 10 Commandments; while I was included in these statistics at one time, we must at least have a good understanding of the faith and the Bible that we profess to know. Moreover, I truly believe that every church that has either Sunday school or Sabbath school should have an apologetics study class. This is critical for any churches in a secular world, (as we discussed in chapter 2). This includes any and all topics that we may be confronted with; evolution, other religions, complexity within scriptures, and the list goes on and on.

Having a simple study is not difficult at all. I literally knew nothing about scripture or apologetics, and in just two short years of prayer and study, I would say that I am at least at an intermediate level. If I can do it, (then trust me), anyone can. So please reference the links and books at the end of the appendix for more information, and please visit my simple site that I threw together, and request a free start-up pack on apologetics.

Now comes the next stage that 100% of us need to be behind, (believer or not). We must insist on our educational systems having an unbiased agenda when it comes to education. What I mean by this is our public schools (k-12) and colleges/universities need to teach the facts to the best of their ability without excuse. Join the school board and insist on evidences of intelligent design be either optional and/or insist on any lies in text books to be removed. (Blatant falsities in favor of evolution are still found in text books, and while evolutionists themselves admit these are false, they do not insist on our educational institutions correcting them, (see Icons of Evolution for more information)). Find out more on your schools' text book selection committee. Companies stand to lose a lot of money if parents and concerned adults begin choosing not to carry their particular books, and they will be only too happy then, to remove any of these falsities. Most teachers, principals, board members, etc., involved in the public school system are sincere, dedicated professionals who want what is best for students. Often they face enormous pressure from small but vocal groups making them feel they are in the battle alone.

> *"The reaction by Darwinists to Jonathan Wells's Icons of Evolution exemplifies this unwillingness to concede anything to intelligent design. In that book, Wells analyzes ten 'icons of evolution.' The reason for calling them icons is that they are presented in high school and college biology textbooks as slam-dunk evidence for evolutionary theory. What has been the response of the evolutionary community? To issue an apology for misleading our young people? To fix the mistakes in the textbooks? No, but to cast Jonathan Wells (who has two doctoral degrees) as a lunatic."* [148]

Another powerful tool that many do not take advantage of, (as mentioned in the above appendix), is contacting your mayor, state representative, governor, white house, and so on; to express your concerns. If you visit www.ACLJ.org you will find a very helpful link that will ask you for your address, and then give you your specific list of all of your representatives, as well as an option to contact each of them with an email, letter, phone call, or set up a meeting. You are their constituency so they have to be open to your needs. If enough of us contact them about the same issues of educational freedoms and for us all to have the ability to take the evidence where it leads, then they will listen. (Texas officials have started to) It is easy, and we can all do this. We can even print petitions and get the signatures ourselves; set-up a meeting with our elected officials and express our concerns. This is a very important factor to consider.

Using the media is another big area. Once again, if I can do this, then anyone can. I paid a small fee and have created several websites. I have received requests from all over the

Christianity in the 22nd Century

world for information on apologetics, Bibles, and more. I also can quickly reference people to this website and ask them to request free information. (If you don't mind spending a little money, then most people won't decline a free offer.) I have rented community college auditoriums (non-profit is $35) and a major university non-profit is $300; to host a guest speaker on intelligent design, apologetics, creation, etc. Once again, if you trust in the Lord, then anything is possible. I have had Dr. Pandit (www.SearchSeminars.org) come and speak twice before, and am planning on a third time in 2010, Lord willing. I also was blessed with Discovery Institute sending me a speaker on intelligent Design for a seminar at the University of Arkansas, February 11th 2010. You will find that these are incredibly effective and many speakers will even take care of their own travel. If I really go all out for an entire seminar cost me about $1,000. If your church or small group would chip in, this literally could be paid for by $25 per person! <u>Think about how many students could benefit from this if 10% of churches in the United States would do this for ONE year!</u>

Form a local pro-ID group. For example, I am a chapter leader of www.ReaonableFaith.org with William Lane Craig. Was this difficult? No, you finish his fantastic book and workbook "Reasonable Faith," and upon completion, you have your own chapter! If you would prefer to stick to the intelligent design piece, visit: www.IdeaCenter.org/clubs to discuss starting a local chapter on intelligent design. Even if you just want to show a debate between a theist such as William Lane Craig, (just type his name in YouTube), or show intelligent design videos, the possibilities are limitless. As mentioned in this book, sitting on the sidelines and waiting for things to get better is no longer an option if you are serious about your faith, the truth, or the future.

In conclusion, do not be discouraged if a lot of your friends or even church is slow to support you. I have had this experience, but once I get over my disappointment, the Lord kicks me in the pants, and I get going again. I know that by me funding 90% of my own events on a $50k salary, I am really limiting myself, so just think what you could do if you can get a group, church, or organization to help spread the finances as well as being prayer partners and encouragers! William Wilberforce decided to go against the slave trade itself. He lost many friends and it was a 20-year-fight in which like-minded people began joining him and he kept his focus on Christ; Wilberforce also started or helped to fund 69 different societies that targeted almost every social and cultural issue of the day, including societies against the use of child labor, medical care for the indigent, British Bible society, missions societies, a national gallery of art, animal rights, and the list goes on. This was ONE individual. **What will you do?**

Sites:

www.IntelligentWonders.com
http://www.rzim.org
www.ReasonableFaith.org
www.discovery.org/csc
www.intelligentdesign.org
www.evolutionnews.org
www.designinference.com
www.intelligentdesignnetwork.org
www.exploreevolution.org
www.ideacenter.org
www.overwhelmingevidence.com
www.dissentfromdarwin.org
www.probe.org

Practical use on how one might actually use evidence/apologetics:

Now many people will ask for just a few basic areas that all believers should be able to agree with, and that are basic, simple, and pretty easy to remember. There are a ton of books on each of these topics that address them in their entirety, (for example – William Lane Craig has an entire book on just the Kalam Cosmological argument); but for the purpose of this book from an average Joe's perspective, we will just touch on some of the basic points, *and how you can simply approach a skeptic or unbeliever. (From my own experience, you will run into some that question God's existence, while some believe in God but have questions about Christianity, so let me give you just a few very broad examples that anyone can easily use, and that the most staunch skeptic has difficulty in refuting even when you or I, (an average Joe), use them). This will at least steer you in the right direction, but I would highly encourage you to either visit www.ReasonableFaith.org or even www.YouTube.com and type in William Lane Craig to view how well these points can work against the most staunch atheist out there. (Most of this will be paraphrased from Dr. William Lane Craig's analysis in his book Reasonable Faith and his countless talks and debates. (I would highly recommend everyone own a copy of his 64 page booklet (God, Are You There?), which provides a summary of his main talking points.

VERY BASICS – STEP ONE IS FOR THE ATHEIST OR ANGNOSTIC:

- *If God does not exist, then Life is ultimately meaningless. As William Lane Craig points out: "If you life is doomed to end in death, then ultimately it does*

not matter how you live." In other words, there is no point to anything. So what if you're rich, if you do good or bad; nothing matters and in the end, there is no ultimate difference whether you existed or not. –This should at least make a person think and reflect on "life" in general. *(pg 4-5 of God, Are You There by William Lane Craig)*
- *On the other hand, though* – if God does exist, then not only does life have meaning and hope, but there is also the possibility of coming to know God and his love personally." *(pg 6 of God, Are You There by William Lane Craig)*
- *Pascal's Wager* – "Therefore, I'm inclined to agree with the French mathematical genius Blaise Pascal that even if the evidence for or against God is equal, the rational thing to do would be to believe in God's existence. That is to say, if the evidence is balanced, then why would someone prefer to bet on death and despair over hope and significance? Therefore, as Craig put it: "I'm inclined to speak of the presumption of theism: we ought to presume that God exists unless we have some good reason to think that atheism is true." *(pg 6-7 of God, Are You There by William Lane Craig)*

At this point, the person you are discussing with should not really be able to say much other than: "Well I prefer despair!" - They are fighting a losing battle in other words. At this point, we should at least be able to encourage them to reflect on the importance and ramifications of their decision and to not take it lightly.

Next is to play a little Q & A and review the evidences:

- Using the most up to date science – we find that the general consensus is that the Universe, space, and time itself did have a beginning, often referred to as "The Big Bang." - While this matches very much with our Genesis account, many will shake off any "biblical talk," but they will admit that the universe had a beginning and that the chances of it popping into existence uncaused is not logical or rational. Therefore, we can use the cosmological argument to surmise the following points:

1. **Whatever begins to exist has a cause.**
2. **The universe began to exist.**
3. **Therefore, the universe has a cause.**

Even the most ardent skeptic should not be able to deny these claims. So we continue the Q&A with concrete facts of science, and thus we are led to our next point on the fact that we are all here by mere "Chance" or it is more than just "chance?"

- **Teleological argument** – Even the most ardent evolutionist and skeptic will admit that the chance of us being here is beyond probability, they will still often argue for chance. *If we use the above points on the sheer numbers that are beyond possibility, we can once again use science and logic to conclude:

1. **The fine-tuning of the universe is due to law, chance, or design.**
2. **It is not due to law or chance.**
3. **Therefore, it is due to design.**

- I often will bring up as briefly as possible, that even if against everything we know in the realm of physics, science, and mathematics, we still have an empty, lifeless world and universe that somehow is perfect in every way for life. *Then touch on yet again, the sheer impossibility of life arranging. (use odds) -
- As briefly as possible discuss the basics of what we now know of DNA – that it is set-up just like the binary code found in computers, but as Bill Gates put it: "Much more complex." How did the original information come about? Could it have been design? (just leave them with this question as well)

1. **Information systems come from a mind that designed them.**
2. **DNA is the product of a massive information system.**
3. **Therefore, DNA is from a mind that designed and input the information code.**

At this point I would remind the person you are talking with that NO ONE is able to answer any of these questions so don't feel bad. The most ardent evolutionist must continue to remind themselves: "We must remember that all of this design just appears to be design."

- The next point I would like to touch on is that of objective moral values. As we mentioned above, objective values are those that are pre-built into us. Murder and rape are wrong. The only way around this, (which I have seen some atheists attempt to do), is to say "there isn't anything truly wrong with murder or rape, we just don't do them." The atheist is putting themselves in an uncomfortable position, because they know deep down that torturing a child and the Nazi genocide of the Jews is truly wrong, and therefore objective moral values do exist, so we can surmise:

1. **If God does not exist, then objective moral values do not exist.**
2. **Objective moral values do exist.**
3. **Therefore, God exists.**

- Very nicely – I prefer to do what many people call the "Columbo Tactic" – where you very nicely as a humble Christian should, ask a few questions, (that you already know the answer to); something that they are sensitive to in particular, as for example the Holocaust: "Do you think if the Nazis had won the war, then the holocaust would have been good?" Hopefully, they will say "NO," and then simply ask them, "if morals are not objective, then why wouldn't it be good?" *This should at least give them something to think about along with the former points, and then you can keep moving down the list, depending on who you're talking to and how they are responding. (everyone is different and will take different amounts of time)

For all intentional purposes, the dialogue up to this point should at least have planted the seed in the most skeptical heart, (Lord willing), to the level of a deist. So how do we bridge the gap from deism to theism? First, thru prayers and always showing ourselves as humble and meek with this knowledge, not arrogant. Next, we speak candidly that if we look at the topics, Christ makes sense of it all. Here are just a couple of examples to get them to at least think and ultimately pray about it; the rest is for the Holy Spirit to do.

At this point I would tie in how Christ makes sense of this "God."

Since there are probably over one million pages on this subject, (and I am an average Joe not commenting on the many other arguments (arguments from history, prophecy, miracles, laws of nature, exceptions to the law of nature, 2nd law of thermodynamics, gradation, the argument of the impossibility of proving the contrary, etc., etc., etc.)), I will let you research most of these for yourself, but here are some ideas on how I like to conclude:

I would begin with a few points –

1. It would make the most sense to begin with researching Christianity because it is testable, historically and experientially.
2. The historic value of such a book is unbelievably accurate in what we know of history, (again – I will let you research this and verify for yourself).
3. No piece of archaeology has ever proved any Biblical statement false; on the contrary, archaeology has done nothing but confirm the Old and New Testaments.
4. Grace is unique to Christianity – all other religions discuss what God "needs" from you. (Does this not seem odd that God would need something from a finite being like us?) - Christianity states that we are all sinners and we are saved solely by God's Grace to us, (through Christ). A few quick examples that you can research for yourself:
   a) Judaism – You are still waiting for the coming Messiah and in the meantime, you adhere to obeying the commandments of God and Moses' laws (to an extent). Also, some laws have been implemented on how many miles you are allowed

to drive on the Sabbath, what things you are able to do and not do, etc. (works based to a degree)

b) Islam – Five pillars of Islam – You are to pray/proclaim that Allah is God and Mohammed is his messenger, pray five times per day facing Mecca, give alms to the poor, fast on Ramadan, and make one pilgrimage to Mecca. (some other groups will list jihad (holy war (some will reference internal war while others say literal war is required)), dress codes, memorizing the Koran, etc., but these are not part of the five pillars. (works based)

c) Hinduism – pantheistic in nature – thousands of Gods, a caste system of peoples that you are born into, (example – Dalits are the poor of the poor and are born into it without ever having the chance to better themselves because they are born into the bottom level of the caste system, (approximately 200 million peoples are in this system.) - An endless cycle of births, deaths, and re-births as you go through an endless cycle of reincarnation.

d) Buddhism – Siddhartha Gautama was a prince that left his home in search of something outside of and more fulfilling than Hinduism. Ultimately is atheistic because God would be substance-less. While there are many different sects within Buddhism, it is mostly focused on being free of feelings and emotions, meditate with the ultimate goal of enlightenment.

3) Then tie in the simplicity of the basic survey of the Old Testament, (which Jew, Christian, and Muslim agree with for the most part), compare the Isaiah scroll of the Dead Sea with what we know of Christ

and ask who they think these passages are referring to? (then tell them our oldest copy is approximately 250 BC)

4) Compare the incarnate spirit that was in the burning bush on Mt. Sinai, the Tabernacle, and King Solomon's Temple, with that of the Temple of Christ.

5) Tie in the birth, life/teachings, death, and resurrection of Jesus, and what it means to all of humankind. The free gift of forgiveness, if we but accept it. *You may want to be careful not to give them the overly liberal view that you can simply say "I accept" and then do whatever you like and continue to sin to your heart's content; this would not be you truly accepting the free gift. (While it is a free gift that is not earned by works, if we truly accept it, we will want to follow Christ and keep God's commandments, (Christ said he has not came to replace the law, but to fulfill); whether someone truly accepts this gift is between them and God; we do not know, (nor are we to judge), but we should at least make certain that it is a free grace well described by Dietrich Bonhoeffer in his book "The Cost of Discipleship," and not a "Do your favorite sin" type of free grace.

After walking the person through these simple steps, the rest is up to the Holy Spirit. I feel very confident that these simple points and trials will stand up against the strictest examination, and I highly encourage you to research these points in detail and polish your points, and be ready to answer the type questions that may be asked of you. (What about evolution, pain & suffering, other religions, etc.) - All we are to do is plant and water and let God grow; as the Bible proclaims "Draw near to me and I will draw near to you," God will take care of the rest. *Then when they come back to ask further questions, we concentrate on the second part of Jesus' command: **"GO – and make disciples – of all nations."**

*Remember – God changes the person, we are just to lead them like the former atheist Richard Wurmbrand was led when he prayed: *"God, I know surely that You do not exist. But if perchance You exist, which I contest, it is not my duty to believe in You; it is Your duty to reveal yourself to me."* (Tortured for Christ page 12) – This simple prayer resulted in Richard Wurmbrand being one of the most influential pastors in the last half of the 20th century and founder of the Voice of the Martyrs. So don't say that "Apologetics" doesn't plant seeds; Richard Wurmbrand spent 14 years in Communist prisons, so he should know. *Christ does the growing; we are told to be prepared to do the planting.

"Draw near to God and He will draw near to you."
– James 4:8

Appendix – C

Authentic Christian faith and what we can do –

I think we all know what "Authentic Faith" is, and that it is 100% focused on Christ. There are legalists (like the Pharisee) that prided themselves in their own piety and good works – Jesus did not support this interpretation. There is also the other extreme, which Dietrich **Bonhoeffer** refers to as "Cheap Grace." This could be those professing to be Christians in word, but not in deed. We usually hear this as either the "prosperity gospel," hypocritical, or non-authentic faith. Many times one will continue in sin or not follow Christ, but they will exclaim: "We're saved by grace! So do whatever you like!" - While we are truly saved by grace, it is improbable that someone who will only follow Christ in word, but not in actuality, has truly accepted His Grace, and free gift. Everyone has to walk their own path, and find out where the Lord is leading them, but authentic faith will show over in thought, word, and deed. As Christ said, we'll carry our own cross in following after Him. Whether it is reaching out to our community, building our personal relationship with Him, supporting ministry opportunities both locally and overseas with at least 1/3 of our income; authentic faith will be reflected in all of these. As one of the most gifted ministers I have ever heard, Pastor Marc Lien repeatedly would end his home church studies and church sermons with the reflective question: "Is Christ in you?" So my humble request is that each day when we look in the mirror, we ask, *"Is Christ in you?"*

Please make sure He is… With no more needing to be said in this appendix other than "Is Christ in you," I will leave you with a few excerpts from Dietrich **Bonhoeffer**'s "Cost of Discipleship" (page numbers noted):

Grace and Discipleship

Cheap grace is the deadly enemy of our Church. We are fighting today for costly grace. Cheap grace means grace **sold on the market** like cheapjacks' wares. The sacraments, the forgiveness of sin, and the consolations of religion are thrown away at cut prices. Grace is represented as the Church's inexhaustible treasury, from which she showers blessings with generous hands, without asking questions or fixing limits. Grace without price; **grace without cost!** The essence of grace, we suppose, is that the account has been paid in advance; and, because it has been paid, everything can be had for nothing.... 45

Cheap grace means grace as a doctrine, a principle, a system. It means forgiveness of sins proclaimed as a general truth, **the love of God taught as the Christian 'conception' of God**. An **intellectual assent** to that idea is held to be of itself sufficient to secure remission of sins.... In such a Church the world finds a cheap covering for its sins; no contrition is required, still less any real desire to be delivered from sin. Cheap grace therefore amounts to a denial of the living Word of God; in fact, a denial of the Incarnation of the Word of God. 45-46

Cheap grace means the justification of sin without the justification of the sinner. Grace alone does everything they say, and so everything can remain as it was before. *'All for sin could not atone.'* **Well, then, let the Christian live like the rest of the world**, let him model himself on the world's standards in every sphere of life, and not presumptuously aspire to live a different life under grace from his old life under sin....

Cheap grace is the grace we bestow on ourselves. Cheap grace is the preaching of forgiveness without requiring repentance, baptism without church discipline, Communion without confession.... Cheap grace is grace without discipleship, **grace without the cross, grace** without Jesus Christ, living and incarnate. 47

Costly grace is the treasure hidden in the field; for the sake of it a man' will gladly go and self all that he has. It is **the pearl of great price** to buy which the merchant will sell all his goods. It is the kingly rule of Christ, for whose sake a man will pluck out the eye which causes him to stumble, it is the call of Jesus Christ at which the disciple leaves his nets and follows him.

Costly grace is the gospel which must be sought again and again and again, the gift which must be asked for, the door at which a man must knock. Such grace is costly because it calls us to follow, and it is grace because it calls us to follow Jesus Christ. It is costly because it costs a man his life, and it is grace because it gives a man the only true life. It is costly because it condemns sin, and grace because it justifies the sinner. Above all, it is costly because it cost God the life of his Son: "ye were bought at a price," and what has cost God much cannot be cheap for us. Above all, it is grace because God did not reckon his Son too dear a price to pay for our life, but delivered him up for us. Costly grace is the Incarnation of God.

Costly grace is the sanctuary of God; it has to be protected from the world, and not thrown to the dogs. It is therefore the living word, the Word of God, which he speaks as it pleases him. Costly grace confronts us as a gracious call to follow Jesus. It comes as a word of forgiveness to the broken spirit and the contrite heart. Grace is costly because it compels a man to submit to the yoke of Christ and follow him; it is grace because Jesus says: "My yoke is easy and my burden is light."

On two separate occasions Peter received the call, "Follow me." It was the first and last word Jesus spoke to his disciple (Mark 1.17; John 21.22). A whole life lies between these two calls. The first occasion was by the lake of Gennesareth, when Peter left his nets and his craft and followed Jesus at his word. The second occasion is when the Risen Lord finds him back again at his old trade. Once again it is by the lake of Gennesareth, and once again the call is: "Follow me." Between the two calls lay a whole life of discipleship in the following of Christ. Half-way between them comes Peter's confession, when he acknowledged Jesus as the Christ of God....48

This grace was certainly not self-bestowed. It was the grace of Christ himself, now prevailing upon the disciple to leave all and follow him, now working in him that confession which to the world must sound like the ultimate blasphemy, now inviting Peter to the supreme fellowship of martyrdom for the Lord he had denied, and thereby forgiving him all his sins. In the life of Peter grace and discipleship are inseparable. He had received the grace which costs. 49

As Christianity spread, and the Church became more secularized, this realization of the costliness of grace gradually faded. The world was Christianized, and grace became its common property. It was to be had at low cost....49

The Call to Discipleship

"The call goes forth, and is at once followed by the response of obedience. .... It displays not the slightest interest in the psychological reason for a man's religious decisions. And why? For the simple reason that the cause behind the immediate following of call by response is Jesus Christ Himself." 61

"Christianity without discipleship is always Christianity without Christ. It remains an abstract idea, a myth which has a place for the Fatherhood of God, but omits Christ as the living Son. ... There is trust in God, but no following of Christ." 64

"He wants to follow, but feels obliged to insist on his own terms to the level of human understanding. The disciple places himself at the Master's disposal, but at the same time retains the right to dictate his own terms. But then discipleship is no longer discipleship, but a program of our own to be arranged to suit ourselves, and to be judged in accordance with the standards of rational ethic." 66

"If we would follow Jesus we must take certain definite steps. The first step, which follows the call, cuts the disciple off from his previous existence. ... The first step places the disciple in the situation where faith is possible. If he refuses to follow and stays behind, he does not learn how to believe." 66-67

Discipleship and the Cross

Jesus Christ must suffer and be rejected. (Mark 8:31-38) This 'must' is inherent in the promise of God—the Scripture must be fulfilled. Here there is a distinction between suffering and rejection. Had He only suffered, Jesus might still have been applauded as the Messiah. 95

Jesus is a rejected Messiah. His rejection robs the passion of its halo of glory. It must be a passion without honor. Suffering and rejection sum up the whole cross of Jesus. To die on the cross means to die despised and rejected of men. Suffering and rejection are laid upon Jesus as a divine necessity, and every attempt to prevent it is the work of the devil, especially when it comes from his own disciples; for it is in fact an attempt to prevent Christ from being Christ. (Peter in Matthew 16)

That shows how t*he very notion of a suffering Messiah was scandal to the Church.* ... Peter's protest displays his own unwillingness to suffer and that means that Satan has gained entry into the Church, and is trying to tear it away from the cross of its Lord.

Jesus must therefore make it clear beyond all doubt that the "must" of suffering applies to his disciples no less than to himself. ... Discipleship means adherence to the person of Jesus, and therefore submission to the law of Christ which is the law of the cross. 96 *[See John 15:20-21]*

When Jesus begins to unfold this inescapable truth to His disciples, He once more sets them free to choose or reject Him. "If any man would come after me," He says. For it is not a matter of course, not even among the disciples. Nobody can be forced, nobody can even be expected to come. He says rather, "If any man" is prepared to spurn all other offers which come his way in order to follow Him. Once again, everything is left for the individual to decide.... 96-97

To deny oneself is to be aware only of Christ and no more of self, to see only Him who goes before and no more the road which is too hard for us. ... All that self-denial can say is: "He leads the way, keep close to Him."

"...and take up his cross." ... Only when we have become completely oblivious of self are we ready to bear the cross for His sake. If in the end we know only Him, if we have ceased to notice the pain of our own cross, we are indeed looking only unto Him. If Jesus had not so graciously prepared us for this word, we should have found it unbearable. 97

To endure the cross is not a tragedy; it is the suffering which is the fruit of an exclusive allegiance to Jesus Christ. When it comes, it is not an accident, but a necessity. ... the suffering which is an essential part of the specifically Christian life.

It is not suffering per se but suffering-and-rejection, and not rejection for any cause of conviction of our own, but rejection for the sake of Christ. If our Christianity has ceased to be serious about discipleship, if we have watered down the gospel into emotional uplift which makes no costly demands and which fails to distinguish between natural and Christian existence, then we cannot help regarding the cross as an ordinary everyday calamity... We have then forgotten that the cross means rejection and shame as well as suffering.

The Psalmist was lamenting that he was despised and rejected of men, and that is an essential quality of the suffering of the cross. But this notion has ceased to be intelligible to a Christianity which can no longer see any difference between an ordinary human life and a life committed to Christ. The cross means sharing the suffering of Christ to the last and to the fullest.

Only a person thus totally committed to discipleship can experience the meaning of the cross. The cross is there, right from he beginning, he has only got to pick it up; there is no need for him to go out and look for a cross for himself... Every Christian has his own cross waiting for him, a cross destined and appointed by God. Each must endure his allotted share of suffering and rejection. 98

But each has a different share: some God deems worthy of the highest form of suffering, and given them the grace of martyrdom, while others He does not allow to be tempted above that they are able to bear....

The cross is laid on every Christian. The first Christ-suffering which every man must experience is the call to abandon the attachments of this world. ... we surrender ourselves to Christ in union with His death—we give over our lives to death. ... When Christ calls a man, He bids him come and die. ...death in Jesus Christ, the death of the old man [or nature] at his call. Jesus' summons to the rich young man was calling him to die, because only the man who is dead to his own will can follow Christ. In fact, every command of Jesus is a call to die, with all our affections and lusts. But we do not want to die...

The call to discipleship... means both death and life... [It] sets the Christian in the middle of the daily arena against sin and the devil. Every day he encounters new temptations, and every day he must suffer anew for Jesus Christ's sake.

The wounds and scars he receives in the fray are living tokens of this participation in the cross of his Lord. 99

But there is another kind of suffering and shame which the Christian is not spared. While ... only the sufferings of Christ are a means of atonement, yet...the Christian also has to undergo temptation [and] bear the sins of others; he too must bear their shame and be driven like a scapegoat from the gates of the city. (Heb. 13:12-15) ...The passion of Christ strengthens him to overcome the sins of others by forgiving them. "Bear ye one another's burdens, and so fulfill the law of Christ. (Gal. 6:2) ...

Suffering then is the badge of true discipleship. The disciple is not above his master... That is why Luther reckoned suffering among the marks of the true Church... If we refuse to take up our cross and submit to suffering and rejection at the hands of men, we forfeit our fellowship with Christ and have ceased to follow Him. But if we lose our lives in His service and carry out cross, we shall find our lives again in the fellowship of the cross with Christ. The opposite of discipleship is to be ashamed of Christ and His cross and all the offense which the cross brings in its train.

Discipleship means allegiance to the suffering Christ... It is a joy and token of His grace. ... Christ transfigures for His own [the early Christian martyrs] the hour of their moral agony by granting them the unspeakable assurance of His presence. In the hour of the cruelest torture they bear for His sake, they are made partakers in the perfect joy and bliss of fellowship with Him. To bear the cross proves to be the only way of triumphing over suffering. ...

Jesus prays to His Father that the cup may pass from Him, and His Father hears His prayer; for the cup of suffering will indeed pass from Him—but only by His drinking it. 101

God speaking to Luther: "Discipleship is not limited to what you can comprehend—it must transcend all comprehension. ... Not to know where you are going is the true knowledge. My comprehension transcends yours. Thus Abraham went forth from His father... not knowing whither he went. ... Behold, that is the way of the cross. You cannot find it yourself, so you must let me lead you as though you were a blind man. Wherefore it is not you, no man... but I myself, who instruct you by my Word and Spirit in the way you should go. Not the work which you choose, not the suffering you devise, but the road which is clean contrary to all you choose or contrive or desire—that is the road you must take. To that I call you and in that you must be my disciple." 103-4

Discipleship and the Individual

"If anyone comes to Me and does not hate his father and mother, wife and children, brothers and sisters, yes, and his own life also, he cannot be My disciple. (Luke 14:26)

Through the call of God, men become individuals... Every man is called separately, and must follow alone. But men are frightened of solitude, and try to protect themselves from it by merging themselves in the society of their fellow-men and in their material environment. They become suddenly aware of their responsibilities and duties, and are loath to part with them. But all this is only a cloak to protect them from having to make a decision. They are unwilling to stand alone before Jesus and to be compelled to decide with their eyes fixed on Him alone.... It is Christ's will that he should be thus isolated, and that he should fix his eyes solely upon him. 105

The Beatitudes

"Blessed are the poor in spirit, for theirs is the kingdom of heaven." Privation is the lot of the disciples in every sphere of their lives. They ... have no security, no possessions to call their own, not even a foot of earth to call their home, no earthly society to claim their absolute allegiance. ... For his sake they have lost all....

The Antichrist also calls the poor blessed, but not for the sake of the cross, which embraces all poverty and transforms it into a source of blessing. He fights the cross with political and sociological ideology. He may call it Christian, but that only makes him a still more dangerous enemy.... 120

"Blessed are they that mourn, for they shall be comforted." With each beatitude the gulf is widened between the disciples and the people, their call to come forth from the people becomes increasingly manifest.... And so the disciples are strangers in the world, unwelcome guests and disturbers of the peace. No wonder the world rejects them! ... they bear their sorrow in the strength of him who bears them up, who bore the whole suffering of the world upon the cross.... The community of strangers find their comfort in the cross, they are comforted by being cast upon the place where the Comforter of Israel awaits them. Thus do they find their true home with their crucified Lord, both here and in eternity. 121-122

"Blessed are the meek: for they shall inherit the earth." This community of strangers possesses no inherent right of its own...nor do they claim such right for they are meek, the renounce every right of their own and live for the sake of Jesus Christ. When reproached, they hold their peace; when treated with violence they endure it patiently; when men drive them from their presence, they yield their ground. They will not go to law to defend their rights nor make a scene when they suffer injustice.

Their right is in the will of their Lord —that and no more. They show by every word and gesture that they do not belong to this earth.... But Jesus says: "they shall inherit the earth." 122-123

"Blessed are they that hunger and thirst after righteousness: for they shall be filled." Not only do the followers of Jesus renounce their rights, they renounce their own righteousness, too. They get no praise for their achievements or sacrifices.... [—for all the praise goes to our King!]

"Blessed are the pure in heart for they shall see God." Who is pure in heart? Only those who have surrendered their hearts completely to Jesus that He may reign in them alone. ...125

"Blessed are the peacemakers: for they shall be called the children of God." The followers of Jesus have been called to peace. When he called them, they found their peace, for **he is their peace**.... But nowhere will that peace be more manifest than where they meet the wicked in peace and are ready to suffer at their hands. The peacemakers will carry the cross with their Lord, for it was on the cross that peace was made. Now that they are partners in Christ's work of reconciliation, they are called the sons of God as he is the Son of God. 125-126

"Blessed are they that have been persecuted for righteousness sake, for theirs is the kingdom of heaven." ...The world will be offended at them and so the disciples will be persecuted for righteousness' sake. Not recognition, but rejection, is the reward they get from the world for their message and works....127

Rejoice and be exceeding glad for great is your reward in heaven; for so persecuted they the prophets which were before you. "For My sake" the disciples are reproached, but because it is for His sake, the reproach falls on Him.

...while Jesus calls them blessed, the world cries: "Away with them, away with them!"

Yes, but whither? To the kingdom of heaven. "Rejoice and be exceeding glad: for great is your reward in heaven." There shall the poor be seen in the halls of joy.... God wipes away the tears from the eyes of those who had mourned upon earth. He feeds the hungry at his Banquet. There stand the scarred bodies of the martyrs, now glorified and clothed in the white robes of eternal righteousness instead of the rags of sin and repentance. The echoes of this joy reach the little flock below as it stands beneath the cross, and they hear Jesus saying: "Blessed are ye!" 128

The Enemy–the "Extraordinary"

There were those who cursed them for undermining the faith and transgressing the law. There were those who hated them of leaving all they had for Jesus' sake.... There were those who persecuted them as prospective dangerous revolutionaries and sought to destroy them. Some of their enemies were numbered among the champions of the popular religion, who resented the exclusive claim of Jesus. 162-163

The Hidden Righteousness

The **life of discipleship** can only be maintained as long as **nothing is allowed to come between Christ and ourselves** – neither the law, nor personal piety, nor even the world. The disciples always look only to their master, never to Christ and the law, Christ and religion, Christ and the world. ... Only by following Christ alone can they preserve a single eye. ... Thus **the heart of the disciple must be set upon Christ alone.** 173-174

(From "Cost of Discipleship" by Dietrich Bonhoeffer **(1906-1945),** *and http://www.crossroad.to/Persecution/ Bonhoeffer.html)*[151]

Appendix – D

Reformation in Islam and what we can do –

While I am not an expert, I have been blessed by learning a great deal from a number of people I have met and talked with. Steven Khoury – a Palestinian Christian in the predominantly Muslim city of Bethlehem, Israel; Brigitte Gabriel, founder of www.ACTforAmerica.org who was persecuted as a Christian in Lebanon and eventually driven out by Muslim forces; my good friend Nouran – who while she is a Muslim, she is more Christian than many self-proclaimed "Christians," (see chapter 3). They and many others all voice the same sentiments: <u>IT IS TIME FOR ISLAM TO REFORM</u>.

It is no lie that 90% of all world violence has the religion of Islam on at least one of the sides. It is not a lie that 150,000+ Christians have been martyred by Islam every year since 1990 for not renouncing their faith. It is not a lie that moderate Muslims are often murdered for not accepting all precepts of Islam or the Koran. It is not a lie that women have basically no rights under Islamic theocracy, (sharia). Much like Christianity's Reformation was a result of Christians seeking to escape the shackles of non-Christian Clergy making up their own laws outside of Scripture or Christ's teaching; it is time for Islam to have its own reformation. A reformation that will have freedom of religion, freedom to follow Christ (if they choose), but at least to have the freedom to.

I often joke with Nouran and ask her: "You drive, you do not wear a head covering, and you have a job, AND you're talking to me (a male). What would happen to you if you were in any country of the Middle East?" She laughs and says *"I would most likely be killed."* Why we are just joking amongst each other, this is indeed a tragic reality. Nouran is

Egyptian and is fortunate enough to have a family that gives her the choice to follow the path she chooses. While they consider themselves devout Muslims and travel to Saudi Arabia, Mecca, etc, Nouran was quite moved by watching the word for word video account of the Gospel of John. When the Jews were seeking to trap Jesus by presenting a woman caught in adultery, and asked Him if she should be stoned according to the law, He of course replies that "let him who is without sin cast the first stone."

Nouran said this is quite touching and has moved her to want to know more about Jesus and His teachings. (I joke (and hope) she does/will consider herself a Messianic Muslim in due time.) There are many just like Nouran who truly want and need this freedom. The freedom to make their own choices and have the evidences before them. If Islam is the right path, then why do the majority of Islamic countries ban or restrict churches, Bibles, internet, etc? Why in strict Islamic countries is converting to another religion punishable by death? As Ravi Zacharias often points out when someone says Islam is the fastest growing religion, he answers no – <u>it is the fastest growing "enforced" religion</u>. I honestly believe if all Islamic countries would allow all the facts to be known for Christianity to be explored with no "anti-conversion" laws, we would see 50% of Muslims leave their faith within a decade, (probably 2/3 would go to Christianity and 1/3 elsewhere).

Again – what can we do? We can pray, we can write, and we can encourage. We can get a free prayer map from www.persecution.com and educate ourselves more as to what we can do. We can also search the web and we'll find a great deal of women and men too, calling for reform in their religion. While our "allies" such as Saudi Arabia are pouring billions of dollars into propagating a strict fascist form of Islam and thus silencing the few moderate Muslims that actually speak out, we can help support the millions (especially women) who are seeking reform; such people as Irshad Manji who wrote "The Trouble With Islam – A Muslim's Call for Reform in Her Faith." (www.muslim-refusenik.com) She points out repeated examples where women are raped, and therefore convicted of adultery and stoned, (for being raped). She pleads for Muslims to speak out against such atrocities and for us to stand with them. Her story is one of many; as Christians or humanists, we owe it to her to write (and insist) on our own governmental officials placing more concern on human rights and religious reforms vs. our own $$$ interests. (www.ACLJ.org)

I will simply summarize her opening letter to the Muslim world, and then you are free to assess it how you will; but you can no longer say you did not know:

"*My Fellow Muslims,*

I have to be honest with you. Islam is on very thin ice with me. I'm hanging on by my fingernails, in anxiety over what's coming next from the self-appointed ambassadors of Allah. When I consider all the fatwas being hurled by the brain trust of our faith, I feel utter embarrassment. Don't you? I hear from a Saudi friend that his country's religious police arrest women for wearing red on Valentine's Day, and I think: Since when does a merciful God outlaw joy – or fun? I read about victims of rape being stoned for 'adultery,' and I wonder how a critical mass of us can stay stone silent.

When non-Muslims beg us to speak up, I hear you gripe that we shouldn't have to explain the behavior of other Muslims. Yet when we're misunderstood, we fail to see that it's precisely because we haven't given people a reason to think differently about us.

On top of that, when I speak publicly about our failings, the very Muslims who detect stereotyping at every turn label me a sell-out. A sell-out to what? To moral clarity? To common decency? To civilization? Yes, I'm blunt. You're just going to have to get used to it. In this letter, I'm asking questions from which we can not longer hide. Why are we all being held hostage by what's happening between the Palestinians and the Israelis? What's with the stubborn streak of anti-Semitism in Islam? Who is the real colonizer of Muslims – America or Arabia? Why are we squandering the talents of women, fully half of God's creation? How can we be so sure that homosexuals deserve ostracism – or death – when the Koran states that everything God made is 'excellent?' Of course, the Koran states more than that, but what's our excuse for reading the Koran literally when it's so contradictory and ambiguous?

Is that a heart attack you're having? Make it fast. Because if we don't speak out against the imperialists within Islam, these guys will walk away with the show. And their path leads to a dead end of more vitriol, more violence, more poverty, more exclusion. Is this the justice we seek for the world that God has leased to us? If it's not, then why don't more of us say so? What I do hear from you is that Muslims are the targets of backlash. In France, Muslims have actually taken an author to court for calling Islam 'the most stupid religion.' Apparently, he's inciting hate. So we assert our rights – something most of us wouldn't have in Islamic countries. But is the French guy wrong to write that Islam needs to grow up? What about the Koran's incitement of hate against the Jews? Shouldn't Muslims who invoke the Koran to justify anti-Semitism be themselves open to a lawsuit? Or would this account to more 'backlash?' What makes us righteous and everybody else racist? Through our screaming self-pity and our conspicuous silences, we Muslims are conspiring against ourselves. We're in crisis, and we're dragging the rest of the world with us. If ever there was a moment for an Islamic reformation, it's now. For the love of God what are we doing about it?

As I view it, the trouble with Islam is that lives are small and lies are big. Totalitarian impulses lurk in mainstream Islam. That's one hell of a charge, I know. Please hear me out. I'll show you what I mean, as calmly as I possibly can..." [149]

May we as bold vs. politically correct as Ms. Manji is; may we also support those in Iran that are yearning for a reformation within Islam; that are desperately seeking to be released from the fascist wall of Islam. Should we find ways to help them break free and have human rights, freedom of religion and the ability to follow Christ if they so choose? Either way, their future is our future. As I have quoted already, world traveler Ravi Zacharias poignantly puts it this way:

"Islam is willing to destroy for the sake of its ideology. I want to suggest that the choice we face is really not between religion and secular atheism, as Sam Harris, Richard Dawkins, Christopher Hitchens, and others have positioned it. Secularism simply does not have the sustaining or moral power to stop Islam. Even now, Europe

> *is demonstrating that its secular worldview cannot stand against the onslaught of Islam and is already in demise. In the end, America's choice will be between Islam and Jesus Christ. History will prove before long the truth of this contention...*"[132]
> - *Ravi Zacharias, The End of Reason, pgs 126-27*

What would Christ say? What would Christ do? Just as Christ was a revolutionary reformer; may we help support Islam's reformers, and in so doing, help un-yoke them from oppression, and at least allow them the freedom of Christ if they choose.

<u>Reaching Muslims for Christ</u>

While there is a plethora of outreach options for reaching the Muslim world, I will direct you to <u>http://www.arabicbible.com/</u> and the below 18 truths from <u>http://www.truthformuslims.com/18_truths.htm</u> that summarizes it nicely. As mentioned in chapter 1 – I will happily send you a ministry start-up package which includes some outreach documentation on reaching our Muslim friends, (<u>www.TheLollards.org</u>). May we act today to change the face of Christianity in the future.

18 Truths About Reaching Muslims in America With the Gospel

Please source <u>http://www.truthformuslims.com/18_truths.htm</u> when quoting from this page.

These 18 statements address the issue of Muslim evangelism in America.

1. Muslims live in virtually every section of the country and there is no reason why every Muslim community cannot be reached with the gospel.

The Apostle Paul traveled to Athens and while there he walked around the city and looked around at the idols. He then began speaking with a group of philosophers and said,

Men of Athens, I observe that you are very religious in all respects. For while I was passing through and examining the objects of your worship, I also found an altar with this inscription, 'TO AN UNKNOWN GOD.' What therefore you worship in ignorance, this I proclaim to you. The God who made the world and all things in it, since He is Lord of heaven and earth, does not dwell in temples made with hands; neither is He served by human hands, as though He needed anything, since He Himself gives to all life and breath and all things. And He made from one, every nation of mankind to live on all the face of the earth, having determined their appointed times, and the boundaries of their habitation, that they should seek God, if perhaps they might grope for Him and find Him, though He is not far from each one of us. (Acts 17:22-27)

Despite all that can be said about U.S. immigration law and the southern border with Mexico and what political and national security policies should be implemented (and much can be said about those critical issues), it is important to note that the Apostle Paul revealed a theological truth about God's role in determining the times and places of where people live.

In other words, God has allowed the United States to be home for millions of Muslims in our generation. The only Muslim community that is more ethnically diverse than the one in America is the temporary community in Mecca, Saudi Arabia during hajj, the annual pilgrimage. So in reality, the most ethnically diverse Muslim community in the world is found in America.

And all of these Muslims need to be reached with the gospel. And God has brought them to a place where we as Christians can reach them with the Gospel. God has given us Christians a great opportunity and a solemn duty to bring the gospel of Jesus Christ to the Muslims of the world right here in America.

Let us seize this opportunity for the honor of Christ and let us fulfill our Christian duty by telling our Muslims friends about the one who died and rose again.

Reaching Muslims for the cause of Christ is an act of the will. It has come upon us – NOW, in our generation and RIGHT HERE, in the United States – to bring the gospel of Jesus Christ to the Muslims.

2. Every obstacle which hinders the proclamation of the gospel among Muslims can be overcome.

There are two kinds of obstacles which may hinder the proclamation of the gospel: spiritual and physical.

First, the spiritual obstacles have already been defeated by the work of Jesus Christ on the cross through the shedding of his blood and his resurrection from death. Spiritual death, spiritual blindness, and evil spirits have all been defeated.

As Christians proclaim the truth of the gospel of Jesus Christ, any and every spiritual obstacle can be overcome because they have already been defeated.

The barriers are spiritual and they are overcome by spiritual means, such as prayer and the Word of God. The spiritual barriers cannot be overcome with evangelistic strategies.

In order to overcome the spiritual obstacles which hinder the proclamation of the Gospel among Muslims, the gospel must be proclaimed to Muslims.

The Apostle Paul wrote:

So faith comes from hearing, and hearing by the word of Christ. (Romans 10:17)

The victory over spiritual obstacles is already won by the risen Christ. So we preach Christ crucified and risen again. And by proclaiming the message, the obstacles which hinder its proclamation are overcome.

In other words, the act of proclaiming the gospel is precisely that which overcomes the spiritual obstacles.

The second set of obstacles are physical. Some of the most difficult physical obstacles include language and cultural barriers. These and all other physical barriers can be overcome. Others include geography and lack of resources, both of which are easily overcome in America.

3. While not all Muslims will come to faith in Christ, all Muslims can be confronted by the message of Christ.

When the message of Christ is proclaimed to Muslims, some will come to faith in Him. Although not all will come to faith, all can be confronted by the message of Christ.

The word confronted is important.

Christians need not be confrontational when proclaiming the message of Christ. But when the message reaches the heart and mind of the hearer, there is a spiritual confrontation with the message.

The message of Christ always confronts death and sin. Even in the cases where people seem to receive new life in Christ under calm, friendly circumstances, there is still a spiritual confrontation going on.

So our message, even in friendship evangelism, confronts those who are spiritually blind and dead.

All Muslims can be confronted by the message of Christ. They will not understand because they are blinded by the god of this world and because they are spiritually dead. But some will understand and receive the message because the Holy Spirit of God draws them to Christ.

Jesus told the story about the seed and the good soil. The seed is the Word of God in this story.

Behold, the sower went out to sow; and as he sowed, some seeds fell beside the road, and the birds came and ate them up.

And others fell upon the rocky places, where they did not have much soil; and immediately they sprang up, because they had no depth of soil.

But when the sun had risen, they were scorched; and because they had no root, they withered away.

And others fell among the thorns, and the thorns came up and choked them out.

And others fell on the good soil, and yielded a crop, some a hundredfold, some sixty, and some thirty.

He who has ears, let him hear. (Matthew 13:3-9)

With all obstacles being overcome and the message proclaimed, there is no reason why every Muslim in America cannot be confronted by the message of Christ. We cannot determine what soil type each person may be, but the seeds can be sowed.

4. The clear commands of Christ compel us to bring the gospel to Muslims, whatever the cost.

Christ is commanding us as his disciples to make disciples of all nations. These are the words of Christ,

Go therefore and make disciples of all the nations, baptizing them in the name of the Father and the Son and the Holy Spirit, teaching them to observe all that I commanded you; and lo, I am with you always, even to the end of the age. (Matthew 28:19-20)

Christ is commanding us to consider the cost of being a disciple as we follow him and make disciples. These are the words of Christ,

Do not think that I came to bring peace on the earth; I did not come to bring peace, but a sword.

For I came to set a man against his father, and a daughter against her mother, and a daughter-in-law against her mother-in-law; and a man's enemies will be the members of his household.

He who loves father or mother more than Me is not worthy of Me; and he who loves son or daughter more than Me is not worthy of Me. And he who does not take his cross and follow after Me is not worthy of Me.

He who has found his life shall lose it, and he who has lost his life for My sake shall find it. (Matthew 10:34-39)

A similar passage in the Gospel of Luke states,

If anyone comes to Me, and does not hate his own father and mother and wife and children and brothers and sisters, yes, and even his own life, he cannot be My disciple. Whoever does not carry his own cross and come after Me cannot be My disciple.

For which one of you, when he wants to build a tower, does not first sit down and calculate the cost, to see if he has enough to complete it? Otherwise, when he has laid a foundation, and is not able to finish, all who observe it begin to ridicule him, saying, "This man began to build and was not able to finish."

Or what king, when he sets out to meet another king in battle, will not first sit down and take counsel whether he is strong enough with ten thousand men to encounter the one coming against him with twenty thousand? Or else, while the other is still far away, he sends a delegation and asks terms of peace.

So therefore, no one of you can be My disciple who does not give up all his own possessions. (Luke 14:26-33)

5. Evangelistic strategies for reaching Muslims in America will be unique because America is unique among all nations.

The Muslim community in America is the most ethnically diverse Muslim community in the world, with one exception.

This exception is the temporary community of Muslim pilgrims from around the world during the annual hajj in Mecca, Saudi Arabia.

America is a place populated by people from every tribe and tongue and nation.

The United States is a place that attracts people from around the world who eventually become Americans.

America is unique because the concept of being an American is unique among the nations.

Evangelistic strategies for reaching Muslims in America will therefore be unique.

Christians do not need to look to overseas models of Muslim evangelism to reach Muslims for Christ in America. Another way of stating this is that Christians in America can discover the best ways to reach Muslims in America by actually doing it.

Reaching Muslims for Christ in America is a wide open door of opportunity for the body of Christ.

There are no constraints from government authorities for Christians to bring the good news of Jesus Christ to the Muslims in America.

This is an exciting, unique, and important work of God in our generation.

6. The core theological teachings of Islam are diametrically opposed to the person and work of Jesus Christ.

Islam denies the deity of Jesus and his death and resurrection.

In the beginning was the Word, and the Word was with God, and the Word was God. He was in the beginning with God. All things came into being by Him, and apart from Him nothing came into being that has come into being. In Him was life, and the life was the light of men. And the light shines in the darkness, and the darkness did not comprehend it. (John 1:1-5)

For He delivered us from the domain of darkness, and transferred us to the kingdom of His beloved Son, in whom we have redemption, the forgiveness of sins. And He is the image of the invisible God, the first-born of all creation. For by Him all things were created, both in the heavens and on earth, visible and invisible, whether thrones or dominions or rulers or authorities— all things have been created by Him and for Him. And He is before all things, and in Him all things hold together. He is also head of the body, the church; and He is the beginning, the first-born from the dead; so that He Himself might come to have first place in everything. For it was the Father's good pleasure for all the fullness to dwell in Him, and through Him to reconcile all things to Himself, having made peace through the blood of His cross; through Him, I say, whether things on earth or things in heaven. (Colossians 1:13-20)

God, after He spoke long ago to the fathers in the prophets in many portions and in many ways, in these last days has spoken to us in His Son, whom He appointed heir of all things, through whom also He made the world. And He is the radiance of His glory and the exact representation of His nature, and upholds all things by the word of His power. When He had made purification of sins, He sat down at the right hand of the Majesty on high; having become as much better than the angels, as He has inherited a more excellent name than they. For to which of the angels did He ever say,

"Thou art My Son, Today I have begotten Thee"? And again, "I will be a Father to Him And He shall be a Son to Me"? And when He again brings the first-born into the world, He says, "And let all the angels of God worship Him." And of the angels He says, "Who makes His angels winds, And His ministers a flame of fire." But of the Son He says, "Thy throne, O God, is forever and ever, And the righteous scepter is the scepter of His kingdom. (Hebrews 1:1-8)

Jesus Christ, the faithful witness, the first-born of the dead, and the ruler of the kings of the earth. To Him who loves us, and released us from our sins by His blood, and He has made us to be a kingdom, priests to His God and Father; to Him be the glory and the dominion forever and ever. Amen. Behold, He is coming with the clouds, and every eye will see Him, even those who pierced Him; and all the tribes of the earth will mourn over Him. Even so. Amen. (Revelation 1:5-7)

Islam denies these truths about Jesus Christ.

Read what the Apostle Paul says about the importance of the resurrection of Jesus Christ.

Now if Christ is preached, that He has been raised from the dead, how do some among you say that there is no resurrection of the dead? But if there is no resurrection of the dead, not even Christ has been raised; and if Christ has not been raised, then our preaching is vain, your faith also is vain. Moreover we are even found to be false witnesses of God, because we witnessed against God that He raised Christ, whom He did not raise, if in fact the dead are not raised. For if the dead are not raised, not even Christ has been raised; and if Christ has not been raised, your faith is worthless; you are still in your sins. (1 Corinthians 15:12-17)

Read what the Apostle John says about the person of Jesus Christ.

Who is the liar but the one who denies that Jesus is the Christ? This is the antichrist, the one who denies the Father and the Son.(1 John 2:22)

By this you know the Spirit of God: every spirit that confesses that Jesus Christ has come in the flesh is from God; and every spirit that does not confess Jesus is not from God; and this is the spirit of the antichrist, of which you have heard that it is coming, and now it is already in the world. (1 John 4:2-3)

For many deceivers have gone out into the world, those who do not acknowledge Jesus Christ as coming in the flesh. This is the deceiver and the antichrist. (2 John 1:7)

Islam denies the person and work of Jesus Christ. This is a rejection of Him as the Son of God, King of Kings, and Lord of Lords and is a rejection of His work on the cross and His resurrection from death.

7. We are bringing the message of Christ to those who are spiritually dead.

Listen to what the Bible says in Ephesians,

And you were dead in your trespasses and sins, in which you formerly walked according to the course of this world, according to the prince of the power of the air, of the spirit that is now working in the sons of disobedience.

Among them we too all formerly lived in the lusts of our flesh, indulging the desires of the flesh and of the mind, and were by nature children of wrath, even as the rest.

But God, being rich in mercy, because of His great love with which He loved us, even when we were dead in our transgressions, made us alive together with Christ (by grace you have been saved), and raised us up with Him, and seated us with Him in the heavenly places, in Christ Jesus, in order that in the ages to come He might show the surpassing riches of His grace in kindness toward us in Christ Jesus.

For by grace you have been saved through faith; and that not of yourselves, it is the gift of God; not as a result of works, that no one should boast. For we are His workmanship, created in Christ Jesus for good works, which God prepared beforehand, that we should walk in them. (Ephesians 2:1-10)

The Bible teaches that everyone without receiving new life in Jesus Christ is spiritually dead. It's important to remember this as we seek to bring the message of Christ to those who have not yet received the free gift of eternal life.

For the wages of sin is death, but the free gift of God is eternal life in Christ Jesus our Lord. (Romans 3:23)

Spiritual truth can only be understood by those who have been made alive spiritually by God.

And even if our gospel is veiled, it is veiled to those who are perishing, in whose case the god of this world has blinded the minds of the unbelieving, that they might not see the light of the gospel of the glory of Christ, who is the image of God. (1 Corinthians 4:3-4)

8. Islam is more than a religion and so our message to Muslims extends beyond the theological differences between Islam and Christianity.

Because Islam is more than a religion, an approach for understanding Islam must go beyond comparing religions.

Comparisons between Islam and Christianity purely on theological grounds are insufficient if we are to gain a full understanding of the nature of Islam and thus be effective in reaching Muslims for Christ long term.

Mohammed created a civilization with laws dealing with every aspect of society and life. Civil laws, religious laws and military tactics are all combined within the Islamic system.

Muslims, in conversations about religion, are comfortable with discussing politics, social injustice, human relations, the roles of men and women in society, and a host of other relevant topics.

Our witness to Muslims, therefore, is often presented in the context of real life issues that meet their needs rather than merely presenting theological truths.

Jesus said,

Come to Me, all who are weary and heavy-laden, and I will give you rest. Take My yoke upon you, and learn from Me, for I am gentle and humble in heart; and you shall find rest for your souls. For My yoke is easy, and My load is light. (Matthew 11:28-30)

Those who follow the Lord Jesus Christ actually experience the great burdens of life lifted from them. This reality — the burdens of life being lifted by Christ, or actually Him giving us the power to rise above the burdens — is a felt need for the average Muslim. This makes the gospel of Christ more relevant than merely giving a presentation of Christian theology.

9. So that people can clearly understand the message of Christ, culture and language should be considered when presenting the gospel.

There are many ethnic groups of people living in the United States who have not heard the message of Christ from their own people or within their own cultural and language.

The movement of these people, from a place in the world where the gospel has not yet been proclaimed widely to the United States, does not mean they are suddenly in a place where they can understand the message of Christ clearly.

In order to bring the message of Christ effectively to these groups of people living in the United States, their culture and language must be considered.

Bibles, New Testaments, and other literature and media in the language of the people being reached is one way of bridging this communication gap.

An understanding of the culture of the people who are intended to receive the message by those bringing the message is another way to reduce misunderstandings that can so easily develop between people of different cultures.

I have learned that the best solution to bringing the message of Christ to people groups in the United States who are isolated linguistically and culturally is to train people in theology, linguistics and cross-cultural communication for the purpose of establishing a group of new followers of Christ who meet together at least weekly for prayer, Bible teaching, fellowship, and sharing the Lord's supper.

This group of new followers of Christ can then bring the message of Christ more effectively to their own people in the years and generations to come.

The Apostle Paul wrote,

To the Jews I became as a Jew, that I might win Jews; to those who are under the Law, as under the Law, though not being myself under the Law, that I might win those who are under the Law; to those who are without law, as without law, though not being without the law of God but under the law of Christ, that I might win those who are without law. To the weak I became weak, that I might win the weak; I have become all things to all men, that I may by all means save some. (1 Corinthians 9:20-22)

Within his ministry, the Apostle Paul adapted himself to the group he was reaching. Paul was equipped to adapt to the cultures of both Jews and non-Jews as he traveled throughout the Roman Empire preaching and teaching the good news of Jesus Christ and establishing groups of new followers of Christ.

For us in America, the task of learning the language and culture of a particular ethnic group requires perseverance. It's not easy but it must be done if we are to effectively communicate the message of Christ in a way that the average Muslim in a particular cultural group can understand.

10. Muslim sensitivities are no excuse for not boldly proclaiming the gospel.

Some Christians are too concerned about negative reactions Muslims may have to the gospel. While it makes no sense to needlessly upset Muslims who are extremely sensitive to anything they perceive to be against Islam, we are compelled to bring the good news of Jesus Christ to all Muslims.

The sensitivities Muslims have about their beliefs should not deter any follower of Christ from obeying the command of Christ to make disciples of all nations.

Muslims are sensitive about their belief in God and the prophets sent by God (and the one they believe is a prophet).

Christians ought to use that sensitivity Muslims have toward spiritual things to boldly proclaim the person and work of Jesus Christ. The gospel is words of life and light! Christians ought to proclaim this good news to Muslims everywhere!

Listen to the words from the gospel which are recorded in the New Testament book written by the Apostle John:

For God so loved the world, that He gave His only begotten Son, that whoever believes in Him should not perish, but have eternal life. For God did not send the Son into the world to judge the world, but that the world should be saved through Him. He who believes in Him is not judged; he who does not believe has been judged already, because he has not believed in the name of the only begotten Son of God. And this is the judgment, that the light is come into the world, and men loved the darkness rather than the light; for their deeds were evil. For everyone who does evil hates the light, and does not come to the light, lest his deeds should be exposed. But he who practices the truth comes to the light, that his deeds may be manifested as having been wrought in God. (John 3:16-21)

Now why should any follower of Christ not boldly bring this wonderful message to Muslims? This is the message of life that will move Muslims from darkness to light!

There is no better way to positively impact the Muslim world than to proclaim this message to Muslims. After all, if they are sensitive about the message it means they are beginning to understand the Truth!

Listen to the words of the Apostle Paul:

For the mystery of lawlessness is already at work; only he who now restrains will do so until he is taken out of the way. And then that lawless one will be revealed whom the Lord will slay with the breath of His mouth and bring to an end by the appearance of His coming; that is, the one whose coming is in accord with the activity of Satan, with all power and signs and false wonders, and with all the deception of wickedness for those who perish, because they did not receive the love of the truth so as to be saved. And for this reason God will send upon them a deluding influence so that they might believe what is false, in order that they all may be judged who did not believe the truth, but took pleasure in wickedness. (2 Thessalonians 2:7-12)

Notice what Paul says about receiving the "love of the truth so as to be saved."

This is why Muslim sensitivities are no excuse for not boldly proclaiming the gospel. The message itself is the way to present Muslims the opportunity to receive the love of the truth so as to be saved! So let's do everything we can to bring the truth to Muslims, it will change the world.

11. Christian sensitivities are no excuse for not boldly proclaiming the gospel.

Some Christians add their own sensitivities to the sensitivities some Muslims have toward the gospel. The result is a weak presentation of our risen Savior. And the sad fact is that some of these Christians are leaders of ministries that are reaching Muslims for Christ.

We shouldn't worry about the sensitivities of Christians when it comes to telling Muslims about Jesus Christ.

12. Making mistakes in bringing the gospel to Muslims is worth the risk because we will never do all things perfectly.

Muslim evangelism is difficult. Throughout history, various evangelistic outreaches to Muslims have had different results. Even the experts cannot teach a perfect method. Fear of making mistakes should never deter us from stepping out in faith to reach Muslims with the gospel. It's never wrong to make mistakes and it's always right to make improvements on our methods. Mistakes provide the path for doing it better over time.

13. Making mistakes in bringing the gospel to Muslims does not close the door of salvation to Muslims who the Holy Spirit is drawing to Christ.

While we should be diligent to do the best we can to present the gospel to Muslims, making mistakes does not diminish the possibility of Muslims coming to Christ. God is in control and the Holy Spirit will draw Muslims to Christ in spite of our weaknesses and failures.

14. Many Muslims are being drawn to Christ by the Holy Spirit in our generation and we must find them and give them the gospel.

More Muslims are coming to Christ now than at any other time in history.

The Holy Spirit is indeed drawing Muslims to Christ. We must find those who are being drawn by the Holy Spirit to give them the gospel. During the process of finding those who are being drawn by the Holy Spirit we will planting seeds in the hearts of those who will listen which will bear fruit in God's timing according to His will.

15. Millions of Christians in America can be mobilized for Muslim ministry.

Millions of Christians in America already have some kind of relationship with a Muslim. Muslims are our neighbors, co-workers, vendors, customers, fellow-students, or friends. It's an easy task for these millions of Christians to live out the life of Christ daily as a witness and to also give something like the Jesus DVD.

16. More full time Christian workers dedicated to bringing the gospel to Muslims are needed to take advantage of the open doors of opportunity in our generation.

Jesus said, "The harvest is plentiful, but the laborers are few; therefore beseech the Lord of the harvest to send out laborers into His harvest" (Luke 10:2). The task before us does require some full-time Christian workers who specialize in Muslim evangelism. Full-time specialists who know the language and the culture of a Muslim people group can train others is basic cross-cultural skills and can lead the way in targeting specific Muslim communities in America.

17. Friendship evangelism is not the only way to bring the gospel to Muslims.

Hospitality and friendship evangelism is a good method for introducing Muslims to Christ, but there are other ways too.

Mass distribution of gospel literature, media, internet, and churches giving public witness in the community can and should be used widely to reach Muslims in America with the gospel.

There is even a place for public debate with Muslims. Often it is the Muslim leaders who challenge Christians to debate and then encourage Muslims to come out and listen to a Christian speaker!

18. The local church is the place to bring former Muslims who believe in Christ.

When Muslims come to faith in Christ, they must get introduced to other Christians. It is excellent when they can be introduced to other former Muslims from their own cultural group for discipleship, fellowship, worship, and spiritual growth. This doesn't have to take place within the walls of a traditional looking church building. This can be small groups and house churches. Whatever the context of the fellowship meetings, they need to get connected with a group of Christians in their local community.

Appendix – E

Creationism – What about the age of the earth and all that?

I wasn't going to comment on this, but since a lot of people will inevitably ask, I thought I would very briefly touch on a few quick points, referencing heavily on the best summary I have seen in Wayne Grudem's book Systematic Theology.[150] So what about the age of the universe/earth? I honestly think this subject takes the focus off of the main subject, but I like Wayne's quote on the subject:

"Although our conclusions are tentative, at this point in our understanding, Scriptures seem to be more easily understood to suggest (but not require) a young earth view, while the observable facts of creationism seem increasingly to favor an old earth view. Both views are possible, but neither one is certain. And we must say very clearly that the age of the earth is a matter that is not directly taught in Scripture, but is something we can think about only by drawing more or less probable inferences from Scripture. Given this situation, it would seem best to admit that God may not allow us to find a clear solution to this question before Christ returns, and to encourage evangelical scientists and theologians who fall in both the young earth and old earth camps to begin to work together with much less arrogance, much more humility, and a much greater sense of cooperation in a common purpose.

There are difficulties with both old earth and young earth viewpoints, difficulties that the proponents of each view often seem unable to see in their own positions. Progress will certainly be made if old earth and young earth scientists who are Christians will be more willing to talk to each other without hostility, ad hominem attacks, or highly emotional accusations, on the one hand, and without a spirit of condescension or academic pride on the other, for these attitudes are not becoming to the body of Christ, nor are they characteristic of the way of wisdom, which is 'first pure, then peaceable, gentle, open to reason, full of mercy and good fruits, without uncertainty or insincerity,' and full of the recognition that 'the harvest of righteousness is sown in peace by those who make peace' (James 3:17-18).

As for evangelism and apologetics done in publications designed to be read outside the evangelical world, young earth and old earth proponents could cooperate much more in amassing the extremely strong arguments for creation by intelligent design, and in laying aside their differences over the age of the earth. Too often young earth proponents have failed to distinguish scientific arguments for creation by design from scientific arguments for a young earth, and have therefore prevented old earth advocates from joining them in battle for the minds of an unbelieving scientific community. Moreover, young earth proponents have sometimes failed to recognize that scientific arguments for a young earth (which seem to them to be very persuasive) are not nearly as strong as the overwhelming scientific arguments for creation by intelligent design. As a result, young earth proponents have too often given the impression that the only true 'creationists' are those who believe not only in creation by God but also in a young earth. The result has been unfortunate divisiveness and lack of community among scientists who are Christians – to the delight of Satan and the grieving of God's Holy Spirit.

Finally, we can view this controversy with some expectancy that there will be further progress in scientific understanding of the age of the earth. It is likely that scientific research in the next ten or twenty years will tip the weight of evidence decisively toward either a young earth or an old earth view, and the weight of Christian scholarly opinion (from both biblical scholars and scientists) will begin to shift decisively in one direction or another. This should not cause alarm to advocates of either position, because the truthfulness of Scripture is not threatened (our interpretations of Genesis 1 have enough uncertainty that either position is possible). Both sides need to grow in knowledge of the truth, even if this means abandoning a long-held position."[150]

Appendix – F

"When I read the book of Acts, I see the church as an unstoppable force. Nothing could thwart what God was doing, just as Jesus foretold: 'The gates of hell shall not prevail against it.' (Matt. 16:18). The church was powerful and spreading like wildfire, not because of clever planning, but by a movement of the Spirit. Riots, torture, poverty, or any other type of persecution couldn't stop it. Isn't that the type of church movement we all long to be a part of? So much of what we see today is anything but unstoppable. It can easily be derailed by the resignation of a pastor or an internal church disagreement or budget cuts. Churches we build only by our own efforts and not in the strength of the Spirit will quickly collapse when we don't push and prod them along. I spent years God to be part of whatever I was doing. When I read the book of Acts, I see people privileged to play a part in what God was doing."[180]

    I am very pleased that you have made it to this point in the book! I am no scholar, I am not financially rich, but I do hope and pray that I am sincere. In so, I would very much like to help you get started in whatever way possible, your own ministry, "Lollard" group, house church, cell group, internet church, or whatever you would like to call it. The heart of the matter (as I have stated throughout the book) is to ask yourself "Is Christ in me?" and then to ask "If He is, then I have got to let others know about Him!" I can offer you a very good start-up package (well over $100) if you go to www.TheLollards.org and request the package. *I will also pay for shipping and handling – all I ask is that you either read, have read, or will read the Bible, and Ravi Zacharias' book

"Beyond Opinion," (which I will have links where you can either read it on-line or can buy it very cheap from Amazon. com); as well as send me back a simple quiz/questionnaire via email that is all on the website. Then I will send you books, tracts, DVDs, CDs, and much more to get you started and on your way! (I do not exaggerate when I say that this will be well worth your time)

Thank you again and God Bless –

www.TheLollards.org

1 Peter 2:9 (New American Standard Bible)

> *[9]But you are a chosen race, a royal priesthood, a holy nation, a people for God's own possession, so that you may proclaim the excellencies of Him who has called you out of darkness into His marvelous light; for you once were not a people, but now you are the people of God; you had not received mercy, but now you have received mercy.*

> "Our Scriptures teach that if you know what you are supposed to do and you don't do it, then you sin (James 4:17). In other words, when we stock up on knowledge without applying it to our lives, we are actually sinning. You would think that learning more about God would be a good thing... and it can be. But when we gain knowledge about God without responding to Him or assimilating His truth into our lives, then it is not a good thing. According to the Bible, it's sin. May we not merely gain knowledge. Instead, as we learn, may we grow and confess and change more into the people we've been created to be by the power of the Holy Spirit, who dwells

within us. "For the kingdom of God is not a matter of eating and drinking but of righteousness and peace and joy in the Holy Spirit" (Rom 14:17).[181]

About the Author

I am really just an average Joe that is incredibly blessed. An average Joe that had a "C" average in high school and a (low) "C" average in college; therefore you should be incredibly encouraged that if I can understand the premises of this book and be able to allow God to use me in whatever humble way I can be used to communicate them to others, then **anyone** can. While I am still learning everyday, I am so thankful that instead of remaining a kid in the shadows that never got out of his comfort zone, I prayed to be used by the Lord in whatever humble way I could be, and I have had quite an adventure in drawing closer to the Lord and better understanding what Christ meant when He said He was *"Truth"* in its essence. I can not stress enough that if I can understand the premises of this book and be able to tell them to others, then anyone reading this can do 10 times as much; the secret is to ask the Lord, and use your talents to the glory of the Lord.

I would encourage anyone wanting to have a basic understanding of the persecuted church to start with the Voice of the Martyrs, and consider doing their on-line certificate workshop, (www.persecution.com). I would also encourage anyone wanting to have a good solid understanding of apologetics to take the apologetics certificate workshop from Biola University. I would also ask you to take the time to read Dietrich Bonhoeffer's book "Cost of Discipleship" along with your opened Bible. These three programs will give you a great foundation to build from, but most importantly I would encourage you to study and know your Bible from inside out, and be in constant prayer for guidance of the Holy Spirit. This will be the most challenging and rewarding endeavor. We must know the faith we profess, and live that faith out in Christ's name.

I am currently an area representative for the Voice of the Martyrs, helping start a Reasonable Faith Chapter (www.ReasonableFaith.org) in Fayetteville Arkansas and it appears that the Lord is pulling me towards New York City where my wife is currently stationed with her job (http://reach-nyc.com). The Lord will lead us if we but ask. Why not ask today?

I wish you all the best. God Bless.
Humbly,
James Stroud
www.IntelligentWonders.com
www.TheLollards.org
www.MartyrsCry.org

> *"Despite everything, I believe that people are really good at heart."*
> Anne Frank

NOTES

1) God, Are You There? William Craig, Ravi Zacharias Intl Ministries (June 1999), pg 4

2) Phillip Jenkins, The Lost History of Christianity – the thousand year old golden age of the church in the Middle East, Africa, and Asia – and how it died, HarperOne; Reprint edition (November 3, 2009), pg 5

3) Phillip Jenkins, The Lost History of Christianity – the thousand year old golden age of the church in the Middle East, Africa, and Asia – and how it died, HarperOne; Reprint edition (November 3, 2009), pg xi

4) Wikipedia, Christianity - http://en.wikipedia.org/wiki/Christianity

5) Robert M. Bowman, Jr. & Kenneth D. Boa, Mission for the Third Millennium Meeting the Challenges to Christian Faith and Values, page 149-150

6) Thomas F. Madden, The New Concise History of the Crusades, Rowman & Littlefield Publishers, Inc.; Stu Upd edition (November 10, 2005), pgs 1-2

7) Brigitte Gabriel, They Must Be Stopped, St. Martin's Griffin (January 5, 2010), pg 27-8

8) Jamsheed Choksy, Conflict and Cooperation, Columbia University Press (July 15, 1997)

9) Ira M. Lapidus, A History of Islamic societies, Cambridge University Press; 2 edition (August 26, 2002)

10) Brigitte Gabriel, They Must Be Stopped, St. Martin's Griffin (January 5, 2010), pg 33

11) Andrew Wheatcroft, Infidels (New York: Penguin Putman, 2003); Richard Fletcher, The Cross and the Crescent (New York: Allen Lane, 2003); A.G. Jamieson, Faith and the Sword (London: Reaktion, 2006); Hugh Kennedy, The Great Arab Conquests (London: Weidenfeld & Nicolson, 2007).

12) Thomas F. Madden, The New Concise History of the Crusades, Rowman & Littlefield Publishers, Inc.; Stu Upd edition (November 10, 2005),, pg 4

13) The Frankish Kingdom. 2001. The Encyclopedia of World History

14) Battle of Tours - Britannica Online Encyclopedia, "Charles's victory has often been regarded as decisive for world history, since it preserved western Europe from Muslim conquest and Islamization."

15) Paul K. Davis, 100 Decisive Battles: From Ancient Times to the Present, Oxford University Press, USA; First Paper edition (June 14, 2001)p. 104.

16) Santosuosso, Anthony . Barbarians, Marauders, And Infidels: The Ways Of Medieval Warfare, Basic Books (May 26, 2004), 2004

17) Poke's Fifteen Decisive Battles

18) An Islamic Europe?, Tomorrow's World, Volume 8, No 3. ; An Islamic Europe?

19) Wikipedia, http://en.wikipedia.org/wiki/Charles_Martel

20) Leonard Ralph Holme, The Extinction of the Christian Churches in North Africa, Nabu Press (February 24, 2010)

21) Phillip Jenkins, The Lost History of Christianity – the thousand year old golden age of the church in the Middle East, Africa, and Asia – and how it died, HarperOne; Reprint edition (November 3, 2009), pg 105

22) Sometimes called the Great Schism, a term that is also applied to the Western Schism (Oxford Dictionary of the Christian Church, Oxford University Press 2005 ISBN 978-0-19-280290-3, article Great Schism)

23) Cross, F. L., ed., ed (2005). The Oxford Dictionary of the Christian Church. New York: Oxford University Press. ISBN 0-19-280290-9.

24) Wikepedia, http://en.wikipedia.org/wiki/East-West_Schism

25) Mingana, Timothy's Apology for Christianity; "thou art empowered" is quoted from Alphonse Mingana, ed.; A Charter of Protection Granted to the Nestorian Church in

AD 1138 (Manchestr, UK: Manchester University Press, 1925)

26) Stephen O'Shea, Sea of Faith (New York: Walker and Company, 2006)

27) Phillip Jenkins, The Lost History of Christianity – the thousand year old golden age of the church in the Middle East, Africa, and Asia – and how it died, HarperOne; Reprint edition (November 3, 2009), page 130

28) Phillip Jenkins, The Lost History of Christianity – the thousand year old golden age of the church in the Middle East, Africa, and Asia – and how it died, HarperOne; Reprint edition (November 3, 2009), page 131

29) Vryonis, Byzantium and Europe, p. 152. "History of the Church Vol II", Innocent III & the Latin East, p372, Philip Hughes, Sheed & Ward, 1948.

30) Sir James Cochran Stevenson Runciman, A History of the Crusades, (pgs 71-102).

31) Hindley, Geoffrey (2004). A Brief History of the Crusades: Islam and Christianity in the Struggle for World Supremacy. London: Constable & Robinson. p. 300. ISBN 978-1-84119-766-1. Wikipedia, http://en.wikipedia.org/wiki/Fall_of_Constantinople

32) Ye'or, Bat. The Decline of Eastern Christianity Under Islam. Fairleigh Dickinson University Press (September 1996), Pg 512

33) Robert M. Bowman, Jr. & Kenneth D. Boa, Mission for the Third Millennium Meeting the Challenges to Christian

Faith and Values, Mission for the Third Millennium— page 144-145

34) Massacre of the Pure, Time, April 28, 1961; European Wars, Tyrants, Rebellions and Massacres (800-1700 CE)

35) Wikipedia, http://en.wikipedia.org/wiki/Persecution_of_Christians

36) Robert M. Bowman, Jr. & Kenneth D. Boa, Mission for the Third Millennium Meeting the Challenges to Christian Faith and Values, Mission for the Third Millennium— page 144-145

37), Brad Steiger, Indian Medicine Power, Whitford Press (November 1984) pgs 79-80

38) Robert M. Bowman, Jr. & Kenneth D. Boa, Mission for the Third Millennium Meeting the Challenges to Christian Faith and Values

39) Frank Tallet, Frank Religion, Society and Politics in France Since 1789 p. 1, 1991 Continuum International Publishing

40) Tallet, Frank Religion, Society and Politics in France Since 1789 p. 1, 1991 Continuum International Publishing, Latreille, A. FRENCH REVOLUTION, New Catholic Encyclopedia v. 5, pp. 972-973 (Second Ed. 2002 Thompson/Gale) ISBN 0-7876-4004-2, SPIELVOGEL, JacksonWestern Civilization: Combined Volume p. 549, 2005 Thomson Wadsworth, and Wikipedia - http://en.wikipedia.org/wiki/Dechristianisation_of_France_during_the_French_Revolution

41) Lewis, Gwynne The French Revolution: Rethinking the Debate p.96 1993 Routledge, ISBN 0415054664 and Wikipedia - http://en.wikipedia.org/wiki/Dechristianisation_of_France_during_the_French_Revolution

42) Tallet, Frank Religion, Society and Politics in France Since 1789 p. 9-11, 1991 Continuum International Publishing and Wikipedia - http://en.wikipedia.org/wiki/Dechristianisation_of_France_during_the_French_Revolution

43) Soviet persecution of Mennonites, 1929-1941 Wikipedia - http://en.wikipedia.org/wiki/Persecution_of_Christians

44) History of the Orthodox Church in the History of Russia Dimitry Pospielovsky 1998 St Vladimir's Press ISBN 0-88141-179-5 pg 291, A History of Marxist-Leninist Atheism and Soviet Antireligious Policies, Dimitry Pospielovsky Palgrave Macmillan (December, 1987) ISBN 0-312-38132-8, Daniel Peris Storming the Heavens: The Soviet League of the Militant Godless Cornell University Press 1998 ISBN 9780801434853, Wikipedia - http://en.wikipedia.org/wiki/Persecution_of_Christians

45) Sermons to young people by Father George Calciu-Dumitreasa. Given at the Chapel of the Romanian Orthodox Church Seminary". The Word online. Bucharest. http://www.orthodoxresearchinstitute.org/resources/sermons/calciu_christ_calling.htm. Wikipedia - http://en.wikipedia.org/wiki/Persecution_of_Christians

46) World Christian trends, AD 30-AD 2200, p.230-246 Tables 4-5 & 4-10 By David B. Barrett, Todd M. Johnson,

Christopher R. Guidry, Peter F. Crossing NOTE: They define 'martyr' on p235 as only including Christians killed for faith and excluding other christians killed, Wikipedia - http://en.wikipedia.org/wiki/Persecution_of_Christians

47) The Washington Post Anti-Communist Priest Gheorghe Calciu-Dumitreasa by Patricia Sullivan Washington Post Staff Writer Sunday, November 26, 2006; Page C09 http://www.washingtonpost.com/wp-, Wikipedia - http://en.wikipedia.org/wiki/Persecution_of_Christians

48) Ostling, Richard (June 24, 2001). "Cross meets Kremlin". TIME Magazine. http://www.time.com/time/magazine/article/0,9171,150718,00.html. Retrieved 2007-07-03.

49) Craughwell, Thomas J., The Gentile Holocaust Catholic Culture, Accessed July 18, 2008, Wikipedia - http://en.wikipedia.org/wiki/Persecution_of_Christians

50) China's Christians suffer for their faith BBC, 9 November 2004. retrieved 25 May 2009, Chinaaid.org, Wikipedia - http://en.wikipedia.org/wiki/Persecution_of_Christians

51) Richard Wurmbrand, Tortured for Christ, Living Sacrifice Book Company (1997) pg 94

52) Dietrich Bonhoeffer, Ethics, Fortress Press (January 2009) pgs 102-109

53) Barrett & Johnson's World Christian Trends A.D. 30-2200: Interpreting the Annual Christian Megacensus, (Pasadena, CA: William Carey Library, 2001), table 4-5, page 230)

54) Richard Wurmbrand, Proofs of God's Existence, Living Sacrifice Book Company pg 122

55) Henkel, Reinhard and Hans Knippenberg "The Changing Religious Landscape of Europe" edited by Knippenberg published by Het Spinhuis, Amsterdam 2005 ISBN 9055892483, pages 7-9, Wikipedia - http://en.wikipedia.org/wiki/Religion_in_europe

56) World Values Survey, Religion and Morals: Believe in God, Wikipedia - http://en.wikipedia.org/wiki/Religion_in_europe

57) Wikipedia - http://en.wikipedia.org/wiki/Religion_in_europe

58) Sam Harris inverview by Bethany Saltman, September 2006 ▪ The Sun

59) Daniel C. Dennett, Darwin's Dangerous Idea: Evolution and the Meaning of Life, Simon & Schuster (June 12, 1996) pg 519

60) Sean McDowell and Willliam Dembski, Understanding Intelligent Design, Harvest House Publishers (July 1, 2008) pg. 14-15

61) Discovery Institute, Center for Science and Culture, Questions about Intelligent Design: What is the theory of intelligent design? Retrieved March 18, 2007. http://www.intelligentdesign.org/, http://www.newworldencyclopedia.org/entry/Intelligent_design

62) Komatsu, E. (2009). "Five-Year Wilkinson Microwave Anisotropy Probe Observations: Cosmological

Interpretation". Astrophysical Journal Supplement 180: 330. doi:10.1088/0067-0049/180/2/330. Bibcode: 2009ApJS..180..330K.; wikipedia - http://en.wikipedia.org/wiki/Big_bang

63) Hubble, E. (1929). "A Relation Between Distance and Radial Velocity Among Extra-Galactic Nebulae". Proceedings of the National Academy of Sciences 15 (3): 168–73. doi:10.1073/pnas.15.3.168. PMID 16577160. PMC 522427. http://antwrp.gsfc.nasa.gov/debate/1996/hub_1929.html., wikipedia - http://en.wikipedia.org/wiki/Big_bang

64) Lemaître, G. (1927). "Un univers homogène de masse constante et de rayon croissant rendant compte de la vitesse radiale des nébuleuses extragalactiques". Annals of the Scientific Society of Brussels 47A: 41. (French) (Translated in: "A Homogeneous Universe of Constant Mass and Growing Radius Accounting for the Radial Velocity of Extragalactic Nebulae". Monthly Notices of the Royal Astronomical Society 91: 483–490. 1931. http://adsabs.harvard.edu/abs/1931MNRAS..91..483L.), wikipedia - http://en.wikipedia.org/wiki/Big_bang

65) Lemaître, G. (1931). "The Evolution of the Universe: Discussion". Nature 128: 699–701. doi:10.1038/128704a0.; wikipedia - http://en.wikipedia.org/wiki/Big_bang

67) God And The Astronomers (1978), W. W. Norton & Company, 2000 2nd edition, paperback: ISBN 0-393-85006-4. The big bang theory and the argument from design; Robert M. Bowman, Jr. & Kenneth D. Boa, Mission for the Third Millennium Meeting the Challenges to Christian Faith and Values, page 36

68) Mark Isaak (ed.) (2005). "CI301: The Anthropic Principle". Index to Creationist Claims. TalkOrigins Archive. http://www.talkorigins.org/indexcc/CI/CI301.html. Retrieved 2007-10-31.

69) Alvin Plantinga, Books & Culture, March/April 2007 Issue

70) William Lane Craig, "The Teleological Argument and the Anthropic Principle"

71) Subodh K Pandit MD, Come Search with Me, Let's Look for God, Xulon Press (June 21, 2008) pg 41

72) Hugh Ross, PhD, The Creator and the Cosmos: How the Greatest Scientific Discoveries of the Century Reveal God, Navpress; 2 edition (July 1995) pg 115

73) http://www.carm.org/apologetics/apologetics/teleological-argument; wikipedia - http://en.wikipedia.org/wiki/Teleology

74) Lee Strobel, The Case for a Creator: A Journalist Investigates Scientific Evidence That Points Toward God, Zondervan (March 1, 2005) pgs 166-67

75) Lee Strobel, The Case for a Creator: A Journalist Investigates Scientific Evidence That Points Toward God, Zondervan (March 1, 2005), pgs 182-3

76) Sylvia Mader, McGraw-Hill Higher Education, Biology, Seventh Edition, Origins of Life pg 320 (ISBN 0-07-118080-X)

77) http://en.wikipedia.org/wiki/Law_of_biogenesis; http://aleph0.clarku.edu/huxley/CE8/B-Ab.html

78) Subodh K Pandit MD, Come Search With Me, Let's Look for God, Xulon Press (June 21, 2008) pg 43

79) William A. Dembski, No Free Lunch: Why Specified Complexity Cannot Be Purchased without Intelligence. (Lanham, Md.: Rowman & Littlefield, 2002.) ISBN 074255810X, p 21-22

80) Robert M. Bowman, Jr. & Kenneth D. Boa, Mission for the Third Millennium Meeting the Challenges to Christian Faith and Values, page 38-40; wikipedia - http://en.wikipedia.org/wiki/Tree_of_life

81) Peter J. Bowler (2003) 'Evolution. The History of an Idea', third edition, p.90-91; wikipedia - http://en.wikipedia.org/wiki/Tree_of_life

82) J. David Archibald (2009) 'Edward Hitchcock's Pre-Darwinian (1840) 'Tree of Life'.', Journal of the History of Biology (2009) 42:561–592; wikipedia - http://en.wikipedia.org/wiki/Tree_of_life

83) wikipedia - http://en.wikipedia.org/wiki/Tree_of_life

84) Sean McDowell and Willliam Dembski, Understanding Intelligent Design, Harvest House Publishers (July 1, 2008) pg. 20-26

85) interview for Philosophia Christi, (vol 6, no 2 pg 201)

86) Michael Behe, PhD, Darwin's Black Box. Free Press, 1996. ISBN 0-684-83493-6; Michael J. Behe, William

A. Dembski and Stephen C. Meyer (2000) Science and Evidence of Design in the Universe. Ignatius Press ISBN 0-89870-809-5

87) William A. Dembski, No Free Lunch: Why Specified Complexity Cannot Be Purchased without Intelligence. (Lanham, Md.: Rowman & Littlefield, 2002.) ISBN 074255810X

88) William A. Dembski, The Design Inference: Eliminating Chance through Small Probabilities. Cambridge: Cambridge University Press, 1998. ISBN 0-521-62387-1

89) William A. Dembski, Intelligent Design: The Bridge between Science and Theology. Downer's Grove, Illinois: InterVarsity Press, 1999. ISBN 0-8308-2314-X

90) William A. Dembski, No Free Lunch: Why Specified Complexity Cannot Be Purchased without Intelligence. (Lanham, Md.: Rowman & Littlefield, 2002.) ISBN 074255810X

91) William A. Dembski, The Design Revolution: Answering the Toughest Questions about Intelligent Design. Downer's Grove, Illinois: InterVarsity Press, 2004. ISBN 0-8308-2375-1

92) Stephen C Meyer, Signature in the Cell: DNA and the Evidence for Intelligent Design, HarperOne (June 23, 2009) HarperCollins

93) PZ Myers: PZ Myers, University of Minnesota Morris, Pharyngula Blog, Zebrafish, Evolutionary Developmental Biology, Cephalopod, Intelligent Design,

Creation?evolution Controversy, Betascript Publishing (January 26, 2010)

94) "Signature in the Cell", Review, Ligonier Ministries; Signature in the Cell Makes 2009 List of Top Ten Best Selling Science Books, Discovery Institute

95) Times Review, The Times.

96) 2009 Books of the Year, The Times

97) Stephen C. Meyer, Signature in the Cell: DNA and the Evidence for Intelligent Design HarperOne (June 23, 2009) ISBN 0061472786; http://en.wikipedia.org/wiki/Steven_C._Meyer

98) Robert M. Bowman, Jr. & Kenneth D. Boa, Mission for the Third Millennium Meeting the Challenges to Christian Faith and Values, page 40

99) www.Discovery.org; www.IntelligentDesign.org

---

100) God and the Astronomers (1978), W. W. Norton & Company, 2000 2nd edition, paperback: ISBN 0-393-85006-4 pgs 11-14

101) Norman L. Geisler & Ronald M. Brooks, When Skeptics Ask: A Handbook on Christian Evidences, pgs 157-59, Baker publishing, 1990 ISBN-13: 978-0801011412

102) Dr. Werner Keller, The Bible as History, Barnes & Noble (December 31, 1995), page xxiii

103) Randall Price, PhD, The Stones Cry Out, Harvest House Publishers (November 1, 1997), page 36-37

104) Norman L. Geisler & Ronald M. Brooks, When Skeptics Ask: A Handbook on Christian Evidences, pgs 179-200, Baker publishing, 1990 ISBN-13: 978-0801011412

105) Josh McDowell, The New Evidence That Demands a Verdict, Thomas Nelson; Rev Upd edition (November 23, 1999) pg 34

106) Ravi Zacharias, "Can Man Live without God?" Thomas Nelson (September 1, 2004), pg 162

107) John Warwick Montgomery PhD, "History, Law, and Christianity," CILTPP (June 30, 2002) pgs 28-36

108) Norman L. Geisler & Ronald M. Brooks, When Skeptics Ask: A Handbook on Christian Evidences, pg 256, Baker publishing, 1990 ISBN-13: 978-0801011412

109) Colin J. Hemer, "The Book of Acts in the Setting of Hellenistic History," Eisenbrauns, 1990

110) (AN Sherwin-White, Roman Society and Roman Law in the New Testament, (Oxford, 1963), 189).

111) Norman L. Geisler & Ronald M. Brooks, When Skeptics Ask: A Handbook on Christian Evidences, pg 223, Baker publishing, 1990 ISBN-13: 978-0801011412

112) Lee Strobel, "The Case for Christ: A Journalist's Personal Investigation of the Evidence for Jesus," Zondervan 1998 ISBN-13: 978-0310209300

113) Debate - Peter Atkins for atheism vs. William Lane Craig for theism

114) William Lane Craig PhD, Reasonable Faith: Christian Truth and Apologetics, Crossway Books; 3 edition 2008 ISBN-13: 978-1433501159

115) Frederick Copleston states in "Problems of Objectivity," in On the History of Philosophy, Image (March 1, 1993) pg 57

116) William Lane Craig PhD, Reasonable Faith: Christian Truth and Apologetics, Crossway Books; 3 edition 2008 pgs ISBN-13: 978-1433501159

117) John Warwick Montgomery PhD, "History, Law, and Christianity," CILTPP (June 30, 2002) pg 22

118) Subodh K Pandit MD, Come Search With Me, Let's Look for God, Xulon Press 2008 (as well as quotes from our meetings)

119) William Lane Craig PhD, "God Are You There," Ravi Zacharias Intl Ministries (June 1999) ISBN-13: 978-1930107007

120) Blaise Pascal, Pensées and Other Writings (Oxford World's Classics), Oxford University Press, USA (July 15, 2008)

121) Alice in Wonderland, Walt Disney Home Video 1951

122) Brad Steiger, "Indian Medicine Power," Whitford Press 1984 pgs 79-80

123) Brad Steiger, "Indian Medicine Power," Whitford Press 1984 pgs 66

124) Christianity Today December 2009, pgs 35 (Joseph Cumming)

125) Michael Frost & Alan Hirsch, "ReJesus: A Wild Messiah for a Missional Church," Hendrickson Publishers (2008), pg 8

126) Steven Khoury, "Diplomatic Christianity," HMC Press (2008) pg 72

127) Dan Kimball, They Like Jesus but Not the Church: Insights from Emerging Generations, Zondervan, pg 104)

128) Dan Kimball, They Like Jesus but Not the Church: Insights from Emerging Generations, Zondervan, pg 74)

129) Dan Kimball, They Like Jesus but Not the Church: Insights from Emerging Generations, Zondervan, pg 48

130) Dan Kimball, They Like Jesus but Not the Church: Insights from Emerging Generations, Zondervan, pgs 218-31)

131) Byzantine Empire, wikipedia - http://en.wikipedia.org/wiki/Byzantine_Empire; The Oxford History of Byzantium, Oxford University Press, USA; First Printing edition (2002)

132) Ravi Zacharias, The End of Reason: A Response to the New Atheists, Zondervan 2008, pgs 126-27

133) William Wilberforce, Real Christianity, Regal Books; Rev Upd edition (2007), pg 87

134) DIETRICH BONHOEFFER, Cost of Discipleship, Touchstone; 1 edition (1995), pgs 43-55

135) William Wilberforce, Real Christianity, Regal Books; Rev Upd edition (2007), pg 59

136) Michael Frost & Alan Hirsch, "ReJesus: A Wild Messiah for a Missional Church," Hendrickson Publishers (2008), pg 79)

137) DC Talk, Jesus Freaks: Revolutionaries: Stories of Revolutionaries Who Changed Their World: Fearing God, Not Man, pg 4

138) Paul Hattaway, "Back To Jerusalem," Gabriel Publishing (2003)

139) John R. W. Scott, The Message of Acts, (Downers Grove; InterVarsity Press, 1994, pg 119

140) Leith Anderson, The Jesus Revolution, Abingdon Press, 2009 pg 55

141) US News and Word Report, Mysteries of the Faith, "In Search of Lost Christian Worlds," Phillip Jenkins, pgs 23-29

142) Johnathen Wells PhD, Icons of Evolution: Science or Myth? Why Much of What We Teach About Evolution is Wrong, Regnery Publishing, Inc.; First Edition,First Trade Paper Edition edition (January 2002) pgs 221-26

143) Kenneth Boa, The Origin of Humanity: The Missing Link (pages 31-33) http://bible.org/byauthor/125/Kenneth%20Boa

144) Paul Hattaway, "Back To Jerusalem," Gabriel Publishing (2003) pgs 14-15

145) Richard Wurmbrand, Tortured for Christ, Living Sacrifice Book Company (1997)

146) Doug Bachelor, Amazing Facts - A Bad Foundation, http://www.amazingfacts.org/Publications/InsideReport/tabid/123/articleType/ArticleView/articleId/432/The-Bible-and-Evolution.aspx

147) Richard N. Ostling, Billy Graham Monday, Nov. 15, 1993, Of Angels, Devils and Messages From God, http://www.time.com/time/magazine/article/0,9171,979587,00.html#ixzz0cWWIbAH3

148) William Dembski and Sean McDowell, Understanding Intelligent Design, Harvest House Publishers 2008, pgs 176-89

149) Irshad Manji, The Trouble with Islam, St Martin's Press NY 2003, pgs 1-3

150) Wayne Grudem, Systematic Theology – An Introduction to Biblical Doctrine, Zondervan 2000, pgs 306-08

151) Dietrich Bonhoeffer, Cost of Discipleship, Touchstone; 1 edition (September 1, 1995), http://www.crossroad.to/Persecution/Bonhoeffer.html

152) Dan Kimball, They Like Jesus but Not the Church: Insights from Emerging Generations, Zondervan, pg 178

153) Paul Hattaway, "Back To Jerusalem," Gabriel Publishing (2003), pgs 64-75, 89

154) Wayne Grudem, Systematic Theology – An Introduction to Biblical Doctrine, Zondervan 2000, pg 135

155) Francis Chan, Forgotten God – Reversing Our Tragic Teglect of the Holy Spirit, Published by David C. Cook, 2009, pg 47.

156) Ayaan Hirsi Ali, Infidel, Free Press – Simon & Schuster, Inc. 2007

157) Joel Richardson, The Islamic Anti-Christ – The Shocking Truth about the Real Nature of the Beast, WordNetDaily publications, Los Angeles, CA, 2009, pgs 4-5

159) Bruce A. McDowell and Anees Zaka, Muslims and Christians at the Table, P & R Publishing (October 1999) pg 26

160) Ravi Zacharias, Jesus Among Other Gods – The Absolute Claims of the Christian Message, Thomas Nelson, Inc., 2000, pgs 158-9

161) Joel Richardson, The Islamic Anti-Christ – The Shocking Truth about the Real Nature of the Beast, WordNetDaily publications, Los Angeles, CA, 2009, pg 213

162) Francis Chan, Forgotten God – Reversing Our Tragic Teglect of the Holy Spirit, Published by David C. Cook, 2009, pg 125-6.

163) Dietrich Bonhoeffer, A Testament to Freedom – The Church is Dead lecture 1932, Harper Collins, 1995, pg. 103.

164) Norm Geisler and Frank Turek, I Don't Have Enough Faith To Be an Atheist, Crossway Books, 2004, pg 102

165) Francis J. Beckwith and Gregory Koukl, Relativism – Feet Firmly Planted in Mid-Air, Baker Books 1998, pg 33

166) Philip Jenkins, The Next Christendom: The Coming of Global Christianity, Oxford University Press, 2002, pg 2

167) Dan Kimball, They Like Jesus but Not the Church: Insights from Emerging Generations, Zondervan, pgs 176-77

168) Rodney Stark, God's Battalions – The Case for the Crusades, HarperOne (September 29, 2009) pg 12

169) Ravi Zacharias, Beyond Opinion – Living the Faith We Defend, Thomas Nelson; Reprint edition (January 12, 2010) pg 255 & 258

170) Steven Khoury, Diplomatic Christianity, HMC Press (2008) pgs 86-87

171) Ravi Zacharias, Beyond Opinion – Living the Faith We Defend, Thomas Nelson; Reprint edition (January 12, 2010) pg 260

172) Richard Wurmbrand, The Triumphant Church, The Voice of the Martyrs (1999) pgs 10-13

173) Ravi Zacharias, Beyond Opinion – Living the Faith We Defend, Thomas Nelson; Reprint edition (January 12, 2010) pg 268-9

174) Derived from Millie Graham Polak's book Gandhi, The Man.by Jyotsna Kamat G. Allen & Unwin (1931)

175) Mahatma Gandhi and Christianity by Dibin Samuel Posted: Thursday, August 14, 2008, 23:50 (IST) Article from Christian Today: http://in.christiantoday.com/article/2837.htm, Copyright (c) 2002 - 2008 Christian Today India

176) Terrence Rynne, Gandhi and Jesus – The Saving Power of Nonviolence, Orbis Books 2008, pg 23.

177) Gandhi, An Autobiography – The Story of My Experiments With Truth, Beacon Press Books 1993, pg xxvi.

178) Ravi Zacharias and Sam Soloman, Beyond Opinion – Living the Faith we Defend, Thomas Nelson; Reprint edition (January 12, 2010) pgs 62 and 69

179) John Julius Norwich, A Short History of Byzantium, First Vintage Books Edition 1999, pg 383

180) Francis Chan, Forgotten God – Reversing our Tragic Neglect of the Holy Spirit, David C. Cook 2009, pg 155

181) Francis Chan, Forgotten God – Reversing our Tragic Neglect of the Holy Spirit, David C. Cook 2009, pg 156-7

182) Ravi Zacharias, Light in the Shadow of Jihad, Multnomah Publishers, 2002, pg 43 and 49

183) Mujahid El Masih & David Witt, Fearless Love – In the Midst of Terror, Martus Publishing 2008 (Faith Covenant International), pg 56

184) Christian History and Biography, Issue 94 Spring 2007, Andrew Saperstein, page 35.